Getting Structured Data from the Internet

Running Web Crawlers/Scrapers on a Big Data Production Scale

Jay M. Patel

Apress®

Getting Structured Data from the Internet: Running Web Crawlers/Scrapers on a Big Data Production Scale

Jay M. Patel
Specrom Analytics
Ahmedabad, India

ISBN-13 (pbk): 978-1-4842-6575-8
https://doi.org/10.1007/978-1-4842-6576-5

ISBN-13 (electronic): 978-1-4842-6576-5

Managing Director, Apress Media LLC: Welmoed Spahr
Acquisitions Editor: Susan McDermott
Development Editor: Laura Berendson
Coordinating Editor: Rita Fernando

Cover designed by eStudioCalamar

Cover image designed by pixabay

Distributed to the book trade worldwide by Springer Science+Business Media New York, 1 New York Plaza, New York, NY 10004. Phone 1-800-SPRINGER, fax (201) 348-4505, e-mail orders-ny@springer-sbm.com, or visit www.springeronline.com. Apress Media, LLC is a California LLC and the sole member (owner) is Springer Science + Business Media Finance Inc (SSBM Finance Inc). SSBM Finance Inc is a **Delaware** corporation.

For information on translations, please e-mail booktranslations@springernature.com; for reprint, paperback, or audio rights, please e-mail bookpermissions@springernature.com.

Apress titles may be purchased in bulk for academic, corporate, or promotional use. eBook versions and licenses are also available for most titles. For more information, reference our Print and eBook Bulk Sales web page at http://www.apress.com/bulk-sales.

Any source code or other supplementary material referenced by the author in this book is available to readers on GitHub via the book's product page, located at www.apress.com/9781484265758. For more detailed information, please visit http://www.apress.com/source-code.

Printed on acid-free paper

To those who believe "Live as if you were to die tomorrow.
Learn as if you were to live forever."

—Mahatma Gandhi.

Table of Contents

About the Author

Jay M. Patel is a software developer with over ten years of experience in data mining, web crawling/scraping, machine learning, and natural language processing (NLP) projects. He is a cofounder and principal data scientist of Specrom Analytics (`www.specrom.com`) providing content, email, social marketing, and social listening products and services using web crawling/scraping and advanced text mining.

Jay worked at the US Environmental Protection Agency (EPA) for five years where he designed workflows to crawl and extract useful insights from hundreds of thousands of documents that were parts of regulatory filings from companies. He also led one of the first research teams within the agency to use Apache Spark–based workflows for chemistry and bioinformatics applications such as chemical similarities and quantitative structure activity relationships. He developed recurrent neural networks and more advanced LSTM models in TensorFlow for chemical SMILES generation.

Jay graduated with a bachelor's degree in engineering from the Institute of Chemical Technology, University of Mumbai, India, and a master of science degree from the University of Georgia, USA.

Jay serves as an editor at a *Medium* publication called Web Data Extraction (`https://medium.com/web-data-extraction`) and also blogs about personal projects, open source packages, and experiences as a startup founder on his personal site (`http://jaympatel.com`).

About the Technical Reviewer

Brian Sacash is a data scientist and Python developer in the Washington, DC area. He helps various organizations discover the best ways to extract value from data. His interests are in the areas of natural language processing, machine learning, big data, and statistical methods. Brian holds a master of science in quantitative analysis from the University of Cincinnati and a bachelor of science in physics from Ohio Northern University.

Acknowledgments

I would like to thank my parents for sparking my interest in computing from a very early age and encouraging it by getting subscriptions and memberships to rather expensive (for us at the time) computing magazines and even buying a pretty powerful PC in summer 2001 when I was just a high school freshman. It served as an excellent platform to code and experiment with stuff, and it was also the first time I coded a basic web crawler after getting inspired by the *ACM Queue*'s search engine issue in 2004.

I would like to thank my former colleagues and friends such as Robbie, Caroline, John, Chenyi, and Gerald and the wider federal communities of practice (CoP) members for stimulating conversations that provided the initial spark for writing this book. At the end of a lot of conversations, one of us would make a remark saying "someone should write a book on that!" Well, after a few years of waiting for that someone, I took the plunge, and although it would've taken four more books to fit all the content on our collective wishlist, I think this one provides a great start to anyone interested in web crawling and natural language processing at scale.

I would like to thank the Common Crawl Foundation for their invaluable contributions to the web crawling community. Specifically, I want to thank Sebastian Nagel for his help and guidance over the years. I would also like to appreciate the efforts of everyone at the Internet Archive, and in particular I would like to thank Gordon Mohr for his invaluable contributions on Gensim listserv.

I am grateful to my employees, contractors, and clients at Specrom Analytics who were very understanding and supportive of this book project in spite of the difficult time we were going through while adapting to the new work routine due to the ongoing Covid-19 pandemic.

This book project would not have come to fruition without the support and guidance of Susan McDermott, Rita Fernando, and Laura Berendson at Apress. I would also like to thank the technical reviewer, Brian Sacash, who helped keep the book laser focused on the key topics.

Introduction

Web scraping, also called web crawling, is defined as a software program or code designed to automate the downloading and parsing of the data from the Web.

Web scraping at scale powers many successful tech startups and businesses, and they have figured out how to efficiently parse terabytes of data to extract a few megabytes of useful insights.

Many people try to distinguish web scraping from web crawling based on the scale of the number of pages fetched and indexed, with the latter being used only when it's done for thousands of web pages. Another point of distinction commonly applied is the level of parsing performed on the web page; web scraping may mean a deeper level of data extraction with more support for JavaScript execution, filling forms, and so on. We will try to stay away from such superficial distinctions and use web scraping and web crawling interchangeably in this book, because our eventual goal is the same: find and extract data in structured format from the Web.

There are no major prerequisites for this book, and the only assumption I have made is that you are proficient in Python 3.x and are somewhat familiar with the SQL language. I suggest that you download and install the Anaconda distribution (`www.anaconda.com/products/individual`) with Python version 3.6.x or higher.

We will take a big picture look in Chapter 1 by exploring how successful businesses around the world and in different domain areas are using web scraping to power their products and services. We'll also illustrate a third-party data source that provides structured data from Reddit and see how we can apply it to gain useful business insights. We will introduce common web crawl datasets and discuss implementations for some of the web scraping applications such as creating an email database like Hunter.io in Chapter 4, a technology profiler tool like builtwith.com, and a website similarity, backlinks, domain authority, and ranking databases like Ahrefs.com, Moz.com, and Alexa.com in Chapters 6 and 7. We will also discuss steps in building a production-ready news sentiments model for alternative financial analysis in Chapter 7.

You will also find that this book is opinionated; and that's a good thing! The last thing you want is a plain vanilla book full of code recipes with no background or opinions on which way is preferable. I hope you are reading this book to learn from the collective experience of others and not make the same mistakes I did when we first started out with crawling the Web over 15 years ago.

I spent a lot of formative years of my professional life working on projects funded by government agencies and giant companies, and the mantra was if it's not built in house, it's trash. Frequently, this aversion against using third-party libraries and publicly available REST APIs is for good reason from a maintainability and security standpoint. So I get it why many companies and new startups prefer to develop everything from scratch, but let me tell you that's a big mistake. The number one rule taught to me by my startup's major investor was: pick your battles, because you can't win them all! He should know, since he was a Vietnam War veteran who ended up having a successful career as a startup investor. Big data is such a huge battlefield, and no one team within a company can hope to ace all the different niches within it except for very few corporations. So based on this philosophy, we will extensively use popular Python libraries such as Gensim, scikit learn, SpaCy for natural language processing (NLP) in Chapter 4, an object-relational mapper called SQLAlchemy in Chapter 5, and Scrapy in Chapter 8.

I think most businesses should rely on cloud infrastructure for their big data workloads as much as possible for faster iteration and quick identification of cost sinks or bottlenecks. Hence, we will extensively talk about a major cloud computing provider, Amazon Web Services (AWS), in Chapter 3 and go through setting up services like IAM, EC2, S3, SQS, and SNS. In Chapter 5, we will cover Amazon Relational Database Service (RDS)–based PostgreSQL, and in Chapter 7, we will discuss Amazon Athena.

You can switch to on-premises data centers once you have documented cost, traffic, uptime percentage, and other parameters. And no, I am not being paid by cloud providers, and for those readers who know my company's technology stack, this is no contradiction. I admit that we run our own servers on premises to handle crawl data, and we also have GPU servers on premises to handle the training of our NLP models. But we have made the decision to go with our setup after doing a detailed cost analysis that included many months of data from our cloud server usage, which conclusively told us about potential cost savings.

I admit that there is some conflict of interest here because my company (Specrom Analytics) is active in the web crawling and data analytics space. So, I will try to keep mentions of any of our products to an absolute minimum, and I will also mention two to three competitors with all my product mentions.

Lastly, let me sound a note of caution and say that scraping/crawling on a big data production scale is not only expensive from the perspective of the number of developer hours required to develop and manage web crawlers, but frequently project managers underestimate the amount of computing and data resources it takes to get data clean enough to be comparable to structured data you get from REST API endpoints.

Therefore, I almost always tell people to look hard and wide for REST APIs from official and third-party data API providers to get the data you need before you think about scraping the same from a website.

If comparable data is available through a provider, then you can dedicate resources to evaluating the quality, update frequency, cost, and so on and see if they meet your business needs. Some commercially available datasets seem incredibly expensive until you factor in computing, storage, and man-hours that go into replicating that in house.

At the very least, you should go out and research the market thoroughly and see what's available off the shelf before you embark on a long web crawling project that can suck time out of your other projects.

CHAPTER 1

Introduction to Web Scraping

In this chapter, you will learn about the common use cases for web scraping. The overall goal of this book is to take raw web crawls and transform them into structured data which can be used for providing actionable insights. We will demonstrate applications of such a structured data from a REST API endpoint by performing sentiment analysis on Reddit comments. Lastly, we will talk about the different steps of the web scraping pipeline and how we are going to explore them in this book.

Who uses web scraping?

Let's go through examples and use cases for web scraping in different industry domains. This is by no means an exhaustive listing, but I have made an effort to provide examples that crawl a handful of websites to those that need crawling a major portion of the visible Internet (web-sized crawls).

Marketing and lead generation

Companies like Hunter.io, Voila Norbert, and FindThatLead run crawlers that index a large portion of the visible Internet, and they extract email addresses, person names, and so on to populate an email marketing and lead generation database. They provide an email address lookup service where a user can enter a domain address and the contacts listed in their database for a lookup fee of $0.0098–$0.049 per contact. As an example, let us enter my personal website's address (jaympatel.com) and see the emails it found on that domain address (see Figure 1-1).

© Jay M. Patel 2020
J. M. Patel, *Getting Structured Data from the Internet*, https://doi.org/10.1007/978-1-4842-6576-5_1

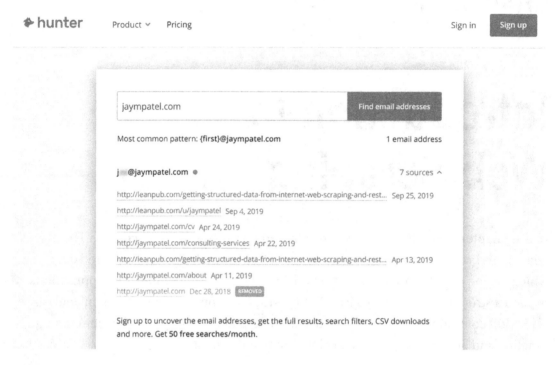

Figure 1-1. *Hunter.io screenshot*

Hunter.io also provides an email finder service where a user can enter the first and last name of a person of interest at a particular domain address, and it can predict the email address for them based on pattern matching (see Figure 1-2).

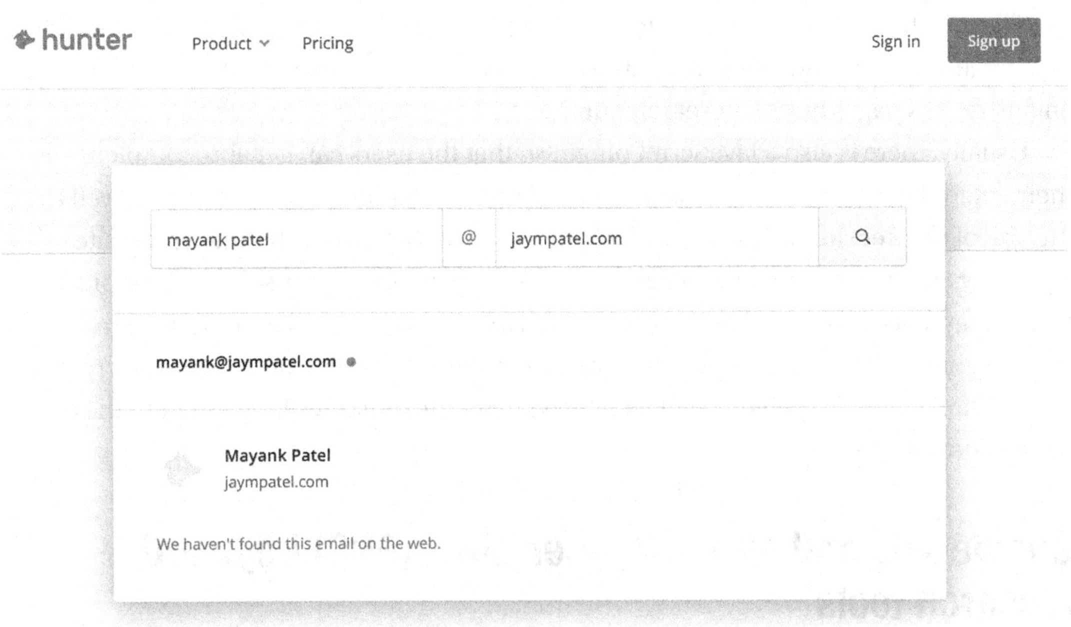

Figure 1-2. *Hunter.io screenshot*

Search engines

General-purpose search engines like Google, Bing, and so on run large-scale web scrapers called web crawlers which go out and grab billions of web pages and index and rank them according to various natural language processing and web graph algorithms, which not only power their core search functionality but also products like Google advertising, Google translate, and so on. I know you may be thinking that you have no plans to start another Google, and that's probably a wise decision, but you should be interested in ranking your business's website higher on Google. This need for being high enough on search engine rankings has spurred off a lot of web scraping/crawling businesses, which I will discuss in the next couple of sections.

On-site search and recommendation

Many websites use third-party providers to power the search box on their website. These are called "on-site searching" in our industry, and some of the SaaS providers are Algolia, Swiftype, and Specrom.

The idea behind all of the on-site searching is simple; they run web crawlers which only target one site, and using algorithms inspired by search engines, they return search engine results pages based on search queries.

Usually, there is also a JavaScript plugin so that the users can get autocomplete for their entered queries. Pricing is usually based on the number of queries sent as well as the size of the website with a range of $20 to as high as $70 a month for a typical site.

Many websites and apps also perform on-site searching in house, and the typical technology stacks are based on Elasticsearch, Apache Solr, or Amazon CloudSearch.

A slightly different product is the content recommendation where the same crawled information is used to power a widget which shows the most similar content to the one on the current page.

Google Ads and other pay-per-click (PPC) keyword research tools

Google Ads is an online advertising platform which predominantly sells ads that are frequently known in the digital marketing field as pay-per-click (PPC) where the advertiser pays for ads based on the number of clicks received on the ads, rather than on the number of times a particular ad is shown, which is known as impressions.

Google, like most PPC advertising platforms, makes money every time a user clicks on one of their ads. Therefore, it's in the best interest of Google to maximize the ratio of clicks per impressions or click-through rate (CTR).

However, businesses make money every time one of those clicked users take an action such as converting into a lead by filling out a form, buying products from your ecommerce store, or personally visiting your brick-and-mortar store or restaurant. This is known as a "conversion." A conversion value is the amount of revenue your business earns from a given conversion.

The real metric advertisers care about is the "return on ad spend" or ROAS which can be defined as the total conversion value divided by your advertising costs. Google makes money based on the number of clicks or impressions, but an advertiser makes money based on conversions. Therefore, it's in your best interest to write ads that don't have a high CTR or click-through rate but rather an ad that has a high conversion rate and high ROAS.

ROAS is completely dependent on keywords, which can be simply defined as words or phrases entered in the search bar of a search engine like Google which triggers your ads. Keywords, or a search query as it is commonly known, will result in a results page consisting of Google Ads, followed by organic results. If we "Google" car insurance, we will see that the top two entries on the results page are Google Ads (see Figure 1-3).

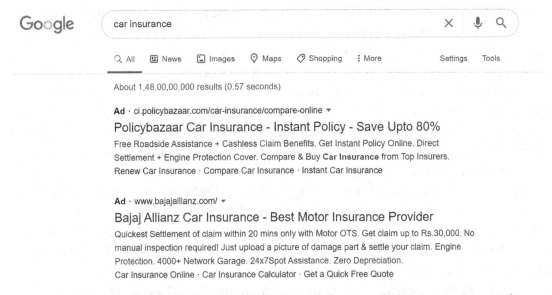

Figure 1-3. *Google Ads screenshot. Google and the Google logo are registered trademarks of Google LLC, used with permission*

If your keywords are too broad, you'll waste a bunch of money on irrelevant clicks. On the other hand, you can block unnecessary user clicks by creating a negative keyword list that excludes your ad being shown when a certain keyword is used as a search query.

This may sound intuitive, but the cost of running an ad on a given keyword on the basis of cost per click (CPC) is directly proportional to what other advertisers are bidding on that keyword. Generally speaking, for transactional keywords, its CPC is directly linked on how much volume of traffic the keyword generates, which in turn drives up its value. If you take an example of transactional keywords for insurance such as "car insurance," the high traffic and the buy intent make its CPC one of the highest in the industry at over $50 per click. There are certain keyword queries made of phrases with two or more words, known as *long tail keywords*, which may actually see lower search traffic but are pretty competitive, and the simple reason for that is that longer keywords with prepositions sometimes capture buyer intent better than just one or two word search queries.

To accurately calculate ROAS, you need a keyword research tool to get accurate data on (1) what others are bidding in your geographical area of interest on a particular keyword, (2) the search volume associated with a particular keyword, (3) keyword suggestions so that you can find additional long tail keywords, and (4) lastly, you would like to generate a negative keyword list that includes words when appearing in a search query do not trigger your ad. As an example, if someone types "free car insurance," that is a signal that they may not buy your car insurance product, and it would be insane to spend $50 on such a click. Hence, you can choose "free" as a negative keyword, and the ad won't be shown to anyone who puts "free" in their search query.

Google's official keyword research tool, called Keyword Planner, included all of the data I listed here up until a few years ago when they decided to change tactics and stopped showing exact search data in favor of insanely broad ranges like 10K–100K. You can get more accurate data if you spend more money on Google Ads; in fact, they don't show any actionable data in the Keyword Planner for new accounts who haven't spent anything on running ad campaigns.

This led to more and more users relying on third-party keyword research providers such as Ahrefs's Keywords Explorer (https://ahrefs.com/keywords-explorer), Ubersuggest (https://neilpatel.com/ubersuggest/), and keywordtool.io/ (https://keywordtool.io/) that provide in-depth keyword research metrics. Not all of them are upfront about their data sourcing methodologies, but an open secret in the industry is that it's coming from extensively scraping data from the official Keyword Planner and supplementing it with clickstream and search query data from a sample population across the world. These datasets are not cheap, with pricing going as high as $300/month based on how many keywords you search. However, this is still worth the price due to unique challenges in scraping Google Keyword Planner and methodological challenges of combining it in such a way to get an accurate search volume snapshot.

Search engine results page (SERP) scrapers

Many businesses want to check if their Google Ads are being correctly shown in a specific geographical area. Some others want SERP rankings for not only their page but their competitor's pages in different geographical areas. Both of these use cases can be easily served by an API service which takes as an input a JSON with a search engine query and geographical area and returns a SERP page as a JSON. There are many providers such as SerpApi, Zenserp, serpstack, and so on, and pricing is around $30 for 5000 searches. From a technical standpoint, this is nothing but adding a proxy IP address, with CAPTCHA solving if required, to a traditional web scraping stack.

Search engine optimization (SEO)

This is a group of techniques whose sole aim is to improve organic rankings on the search engine results pages (SERPs).

There are dozens of books on SEO and even more blog posts, all describing how to improve your SERP ranking; we'll restrict our discussions on SEO here to only those factors which directly need web scraping.

Each search engine uses their own proprietary algorithm to determine rankings, but essentially the main factors are relevance, trust, and authority. Let us go through them in greater detail.

Relevance

These are group of factors that measure how relevant a particular page is for a given search query. You can influence the ranking for a set of keywords by including them on your page and within meta tags on your page.

Search engines rely on HTML tags called "meta" to enable sites such as Google, Facebook, and Twitter to easily find certain information not visible to normal web users. Web masters are not mandated to insert these tags at all; however, doing so will not only help users on search engine and social media find information, but that will increase your search rankings too.

You can see these tags by right-clicking any page in your browser and clicking " view source. " As an example, let us get the source from Quandl.com; you may not yet be familiar with this website, but the information in the meta tags (meta property= " og:description and meta name= " twitter:description) tells you that it is a website for datasets in the financial domain (see Figure 1-4).

```
<meta http-equiv="Content-Type" content="text/html; charset=UTF-8"/>
<meta http-equiv="X-UA-Compatible" content="IE=edge,chrome=1"/>

<meta name="viewport" content="width=device-width, initial-scale=1, maximum-scale=1, user-scalable=yes"/>

<meta name="robots" content="index, follow"/>

<link rel="search" type="application/opensearchdescription+xml" title="Quandl" href="/opensearch.xml"/>

<link rel="apple-touch-icon-precomposed" href="/images/head/apple-touch-icon.png">
<link rel="icon" href="/images/head/apple-touch-icon.png">
<link rel="image_src" type="image/png" href="/images/head/quandl-logo-social.png">

<meta property="og:title" content="Quandl">
<meta property="og:url" content="https://www.quandl.com">
<meta property="og:description" content="The source for financial, economic, and alternative datasets, serving investment professionals.">
<meta property="og:image" content="http://www.quandl.com/images/head/quandl-logo-social.png">
<meta property="og:type" content="website">

<meta name="twitter:card" content="summary">
<meta name="twitter:image" content="/images/head/quandl-logo-twitter.png">
<meta name="twitter:site" content="@quandl">
<meta name="twitter:title" content="Quandl">
<meta name="twitter:description" content="The source for financial, economic, and alternative datasets, serving investment professionals.">

<meta name="google-site-verification" content="9qvd1SCSyNwmz9cxGqxgtXJHqwsWkVkYUyuz0o5SRys"/>
<meta name="msvalidate.01" content="9AA8D9E90DBCA4138E42EF9D32B42C01"/>
```

Figure 1-4. *Meta tags*

It's pretty easy to create a crawler to scrape your own website pages and see how effective your *on-page optimization* is so that search engines can " find " all the information and index it on their servers. Alternately, it's also a good idea to scrape pages of your competitors and see what kind of text they have put in their meta tags. There are countless third-party providers offering a " freemium" audit report on your on-page optimization such as `https://seositecheckup.com`, `https://sitechecker.pro`, and `www.woorank.com/`.

Trust and authority

Obtaining a high relevance score to a given search query is important, but not the only factor determining your SERP rankings. The other factor in determining the quality of your site is how many other high-quality pages link to your site's page (backlinks). The classic algorithm used at Google is called PageRank, and now even though there are a lot of other factors that go into determining SERP rankings, one of the best ways to rank higher is get backlinks from other high-quality pages; you will hear a lot of SEO firms call this the "link juice," which in simple terms means the benefit passed on to a site by a hyperlink.

In the early days of SEO, people used to try "black hat" techniques of manipulating these rankings by leaving a lot of spam links to their website on comment boxes, forums, and other user-generated contents on high-quality websites. This rampant gaming of the system was mitigated by something known as a "nofollow" backlink, which basically meant that a webmaster could mark certain outgoing links as "nofollow" and then no link juice will pass from the high-quality site to yours. Nowadays, all outgoing hyperlinks on popular user-generated content sites like Wikipedia are marked with "nofollow," and thankfully this has stopped the spam deluge of the 2000s. We show an example in Figure 1-5 of an external nofollow hyperlink at the Wikipedia page on PageRank; don't worry about all the HTML tags, just focus on the <a rel = "nofollow" for now.

```
▼<cite class="citation book">
    "Bradley C. Love & Steven A. Sloman. "
    <a rel="nofollow" class="external text" href="http://bradlove.org/papers/
    love_sloman_1995.pdf">"Mutability and the determinants of conceptual
    transformability"</a> == $0
    <span class="cs1-format">(PDF)</span>
    ". "
```

Figure 1-5. *Nofollow HTML links*

Building backlinks is a constant process because if you aren't ahead of your competitors, you can start losing your SERP ranking. Alternately, if you know your competitor's site's backlinks, then you can target those websites by writing compelling content and see if you can "steal" some of the backlinks to boost your SERP rankings. Indeed, all of the strategies I mention here are followed by top SEO agencies every day for their clients.

Not all backlinks are gold. If your site gets disproportionate amount of backlinks from low-quality sites or spam farms (or link farms as they are also known), your site will also be considered "spammy," and search engines will penalize you by dropping your ranking on SERP. There are some black hat SEOs out there that rapidly take down rankings of their competitor's sites by using this strategy. Thankfully, you can mitigate the damage if you identify this in time and disavow those backlinks through Google Search Console.

Until now, I think I have made the case about why it's useful to know your site's backlinks and how people will be willing to pay if you can give them a database where they can simply enter either their site's URL or their competitors and get all the backlinks.

Unfortunately, the only way to get all the backlinks is by crawling large portions of the Internet, just like search engines do, and that's cost prohibitive for most businesses or SEO agencies to do in themselves. However, there are a handful of companies such as Ahrefs and Moz that operate in this area. The database size for Ahrefs is about 10 PB

(= 10,000 TB) according to their information page (`https://ahrefs.com/big-data`); the storage cost alone for this on Amazon Web Services (AWS) S3 would come out to over $200,000/month so it's no surprise that subscribing to this database is pricey at cheapest licenses starting at hundreds of dollars a month.

There is a free trial to the backlinks database which can be accessed here (`https://ahrefs.com/backlink-checker`); let us run an analysis on apress.com.

Figure 1-6. *Ahrefs screenshot*

We see that Apress has over 1,500,000 pages linking back to it from about 9500 domains, and majority of these backlinks are "dofollow" links that pass on the link juice to Apress. The other metric of interest is the domain rating (DR), which normalizes a given website's backlink performance on a 1–100 scale; the higher the DR score, the more "link juice" passed from the target site with each backlink. If you look at Figure 1-6, the top backlink is from `www.oracle.com` with its DR being 92. This indicates that the page is of highest quality, and getting such a top backlink helped Apress's own DR immensely, which drove traffic to its pages and increased its SERP rankings.

Estimating traffic to a site

Every website owner can install analytics tools such as Google Analytics and find out what kind of traffic their site gets, but you can also estimate traffic by getting a domain ranking based on backlinks and performing some clever algorithmic tricks. This is indeed what Alexa does, and apart from offering backlink and keyword research ideas, they also give pretty accurate site traffic estimates for almost all websites. Their service is pretty pricey too, with individual licenses starting at $149/month, but the underlying value of their data makes this price tag reasonable for a lot of folks. Let us query Alexa for apress.com and see what kind of information it has collected for it (see Figure 1-7).

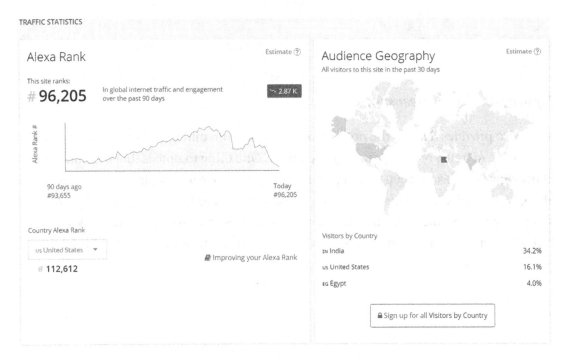

Figure 1-7. *Alexa screenshot*

Their web-crawled database also provides a list of similar sites by audience overlap which seems pretty accurate since it mentions manning.com (another tech publisher) with a strong overlap score (see Figure 1-8).

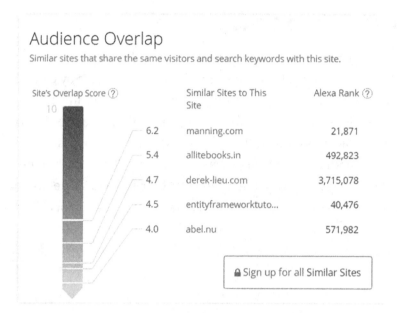

Figure 1-8. *Alexa screenshot*

It also provides data on the number of backlinks from different domain names and percentage of traffic received via search engines. One thing to note is that the number of backlinks by Alexa is 1600 (see Figure 1-9), whereas the Ahrefs database mentioned about 9000. Such discrepancies are common among different providers, and that just shows you the completeness of web crawls each of these companies is undertaking. If you have a paid subscription to them, then you can get the entire list and check for omissions yourself.

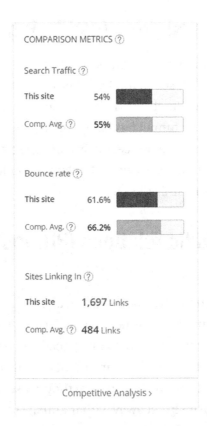

Figure 1-9. Alexa screenshot showing the number of backlinks

Vertical search engines for recruitment, real estate, and travel

Websites such as indeed.com, Expedia, and Kayak all run web scrapers/crawlers to gather data focusing on specific segment of online content which they process further to extract out more relevant information such as name of the company, city, state, and job title in the case of indeed.com, which can be used for filtering through the search results. The same is true of all search engines where web scraping is at the core of their product, and the only differentiation between them is the segment they operate in and the algorithms they use to process the HTML content to extract out content which is used to power the search filters.

Brand, competitor, and price monitoring

Web scraping is used by companies to monitor prices of various products on ecommerce sites as well as customer reviews, social media posts, and news articles for not just their own brands but also for their competitors. This data helps companies understand how effective their current marketing funnel has been and also lets them get ahead of any negative reviews before they cause a noticeable impact on sales. There are far too many examples in this category, but Jungle Scout, AMZAlert, AMZFinder, camelcamelcamel, and Keepa all serve a segment of this market.

Social listening, public relations (PR) tools, and media contacts database

Businesses are very interested in what their existing and potential customers are saying about them on social media websites such as Twitter, Facebook, and Reddit as well as personal blogs and niche web forums for specialized products. This data helps businesses understand how effective their current marketing funnel has been and also lets them get ahead of any negative reviews before they cause a noticeable impact on sales. Small businesses can usually get away with manually searching through these sites; however, that becomes pretty difficult for businesses with thousands of products on ecommerce sites. In such cases, they use professional tools such as Mention, Hootsuite, and Specrom, which can allow them to do bulk monitoring. Almost all of these get some fraction of data through web crawling.

In a slightly different use case, businesses also want to guide their PR efforts by querying for contact details for a small number of relevant journalists and influencers who have a good following and readership in a particular niche. The raw database remains the same as previously discussed, but in this case, the content is segmented by topics such as apparels, fashion accessories, electronics, restaurants, and so on and results combined with a contacts database. A user should be able to query something like find email addresses and phone numbers for ten top journalists/influencers active in the food, beverage, and restaurant market in the Pittsburgh, PA area. There are too many products out there, but some of them include Muck Rack, Specrom, Meltwater, and Cision.

Historical news databases

There is a huge demand out there for searching historical news articles by keyword and returning news titles, content body, author names, and so on in bulk to be used for competitor, regulatory, and brand monitoring. Google News allows a user to do it to some extent, but it still doesn't quite meet the needs of this market. Aylien, Specrom Analytics, and Bing News all provide an API to programmatically access news databases, which index 10,000–30,000 sources in all major languages in near real time and archives going back at least five or more years. For some use cases, consumers want these APIs coupled to an alert system where they get automatically notified when a certain keyword is found in the news, and in those cases, these products do cross over to social listening tools described earlier.

Web technology database

Businesses want to know about all the individual tools, plugins, and software libraries which are powering individual websites. Of particular interest is knowing about what percentage of major sites run a particular plugin and if that number is stable, increasing, or decreasing.

Once you know this, there are many ways to benefit from it. For example, if you are selling a web plugin, then you can identify your competitors, their market penetration, and use their customers as potential leads for your business.

All of the data I mentioned here can be aggregated by web crawling through millions of websites and aggregating the data in headers and response by a plugin type or displaying all plugins and tools used by a certain website. Examples include BuiltWith and SimilarTech, and basic product offerings start at around $290/month with prices going as high as a few thousand a month for searching unlimited websites/plugins.

Alternative financial datasets

Any company-specific datasets published by third-party providers consisting of data compiled and curated from nontraditional financial market sources such as social/sentiment data and social listening, web scraping, satellite imagery, geolocation to measure foot traffic, credit card transactions, online browsing data, and so on can be defined as alternative financial datasets.

These datasets are mainly used by quantitative traders or algorithmic traders who can be simply defined as traders engaged in buying/selling of securities on stock exchanges solely on the basis of computer algorithms. Now these so-called algorithms or trading strategies are rule based and coded by traders themselves, but the actual buy/sell triggers happen automatically once the strategy is put into production.

A handful of hedge funds started out with quantitative trading over 10 years ago and consuming alternative datasets that provided trading signals or triggers powering their trading strategies. Now, however, almost all institutional investors in the stock market from small family offices to large discretionary funds use alternative datasets to some extent.

A large majority of alternative datasets are created by applying NLP algorithms for sentiments, text classification, text summarization, named entity recognition, and so on on web crawl data described in earlier sections, and therefore this is becoming a major revenue stream for most big data and data analytics firms including Specrom Analytics.

You can explore all kinds of available alternative datasets on marketplaces such as Quandl, which has data samples for all the popular datasets such as web news sentiments (`www.quandl.com/databases/NS1`) for more than 40,000 stocks.

Miscellaneous uses

There are a lot of use cases that are hard to define and put into one of these distinct categories. In those cases, there are businesses that offer data on demand, with the ability to convert any website data into an API. Examples include Octoparse, ParseHub, Webhose.io, Diffbot, Apify, Import.io, Dashblock, and so on. There are other use cases such as security research, identity theft monitoring and protection, plagiarism detection, and so on—all of which rely on web-sized crawls.

Programmatically searching user comments in Reddit

Let's work through an example to search through all the comments in a subreddit by accessing a free third-party database called pushshift.io and perform sentiment analysis on it by using algorithms on the request service at Algorithmia.

Aggregating sentiments from social media, news, forums, and so on represents a very common use case in alternative financial datasets, and here we are trying to just get a taste for it by doing it on one major company.

You will also learn how to communicate with web servers using the Hypertext Transfer Protocol (HTTP) methods such as GET and POST requests with authentication, which will be useful throughout this book, as there can be no web scraping/crawling without fetching the web page.

Reddit provides an official API, but there are a lot of limitations to its use compared to pushshift which has compiled the same data and made it available either through an API (`https://github.com/pushshift/api`) or through raw data dumps (`https://files.pushshift.io/reddit/`).

We will use the Python requests package to make GET calls in Python 3.x; it's much more intuitive than the urllib in the Python standard library.

The request query is pretty simple to understand. We are searching for the keyword "Exxon" in the top stock market–related subreddit called "investing" which has about one million subscribers (see Listing 1-1). We are restricting ourselves to a maximum of 100 results and searching between August 20, 2019, and December 10, 2019, so that the request doesn't get timed out. Users are encouraged to go through the pushshift.io documentation (`https://github.com/pushshift/api`) and generate their own query as a learning exercise. The time used in the query is epoch time which has to be converted to date or vice versa by using an online calculator (`www.epochconverter.com/`) or pd.to_datetime().

Listing 1-1. Calling the pushshift.io API

```python
import requests
import json

test_url = 'https://api.pushshift.io/reddit/search/comment/?q=Exxon&subredd
it=investing&size=100&after=1566302399&before=1575979199&sort=asc&metadata=
True'

r = requests.get(url = test_url)

print("Status Code: ", r.status_code)
print("*"*20)
print(r.headers)

html_response = r.text
```

Output

```
Status Code:   200
*******************
```

```
{'Date': 'Wed, 15 Apr 2020 11:47:37 GMT', 'Content-Type': 'application/
json; charset=UTF-8', 'Transfer-Encoding': 'chunked', 'Connection':
'keep-alive', 'Set-Cookie': '__cfduid=db18690163f5c909d973f1a67bb
dc79721586951257; expires=Fri, 15-May-20 11:47:37 GMT; path=/; domain=.
pushshift.io; HttpOnly; SameSite=Lax', 'cache-control': 'public, max-
age=1, s-maxage=1', 'Access-Control-Allow-Origin': '*', 'CF-Cache-Status':
'EXPIRED', 'Expect-CT': 'max-age=604800, report-uri="https://report-uri.
cloudflare.com/cdn-cgi/beacon/expect-ct"', 'Vary': 'Accept-Encoding',
'Server': 'cloudflare', 'CF-RAY': '58456ecf7ee0e3ce-ATL', 'Content-
Encoding': 'gzip', 'cf-request-id': '021f4395ae0000e3ce5d928200000001'}
```

We see that the response code was 200, meaning that the request has succeeded and the response content-type is application/json. We'll use the JSON package to read and save the raw response (see Listing 1-2).

Listing 1-2. Parsing a JSON response

```
with open("raw_pushshift_response.json", "w") as outfile:
    outfile.write(html_response)

json_dict = json.loads(html_response)
json_dict.keys()

json_dict["metadata"]
```

```
# output
{'after': 1566302399,
 'agg_size': 100,
 'api_version': '3.0',
 'before': 1575979199,
 'es_query': {'query': {'bool': {'filter': {'bool': {'must': [{'terms':
{'subreddit': ['investing']}},
      {'range': {'created_utc': {'gt': 1566302399}}},
      {'range': {'created_utc': {'lt': 1575979199}}},
```

```
      {'simple_query_string': {'default_operator': 'and',
        'fields': ['body'],
        'query': 'Exxon'}}],
     'should': []}},
  'must_not': []}},
 'size': 100,
 'sort': {'created_utc': 'asc'}},
'execution_time_milliseconds': 31.02,
'index': 'rc_delta2',
'metadata': 'True',
'q': 'Exxon',
'ranges': [{'range': {'created_utc': {'gt': 1566302399}}},
 {'range': {'created_utc': {'lt': 1575979199}}}],
'results_returned': 71,
'shards': {'failed': 0, 'skipped': 0, 'successful': 4, 'total': 4},
'size': 100,
'sort': 'asc',
'sort_type': 'created_utc',
'subreddit': ['investing'],
'timed_out': False,
'total_results': 71}
```

We see that we only got back 71 results out of a maximum request of 100.

Let us explore the first element in our data list to see what kind of data response we are getting back (see Listing 1-3).

Listing 1-3. Viewing JSON data

```
json_dict["data"][0]
```

```
Output:
{'all_awardings': [],
 'author': 'InquisitorCOC',
 'author_flair_background_color': None,
 'author_flair_css_class': None,
 'author_flair_richtext': [],
```

```
 'author_flair_template_id': None,
 'author_flair_text': None,
 'author_flair_text_color': None,
 'author_flair_type': 'text',
 'author_fullname': 't2_mesjk',
 'author_patreon_flair': False,
 'body': 'Individual stocks:\n\nBoeing and Lockheed: initially languished
until 1974, then really took off and gained almost 100x by the end of the
decade.\n\nHewlett-Packard: volatile, but generally a consistent winner
throughout the decade, gained 15x.\n\nIntel: crashed >70% during the worst
of 1974, but bounced back very quickly and went on to be a multi bagger.\n\
nOil stocks had done of course very well, Halliburton and Schlumberger were
the low risk, low volatility, huge gain stocks of the decade. Exxon on the
other hand had performed nowhere as well as these two.\n\nWashington Post:
fought Nixon head on in 1973, stocks dropped big. More union troubles in
1975, but took off afterwards. Gained between 70x and 100x until 1982.\n\
nOne cannot mention WaPo without mentioning Berkshire Hathaway. Buffett
bought 10% in 1973, got himself elected to its board, and had been advising
Cathy Graham. However, BRK was a very obscure and thinly traded stock back
then, investors would have a hard time noticing it. Buffett himself said
the annual meeting in 1978 all fit in one small cafeteria.\n\n\n\nOther
asset classes:\n\nCommodities in general had performed exceedingly well.
Gold went from 35 in 1970 all the way to 800 in 1980.\n\nReal Estate had
done well. Those who had the foresight to buy in SF Bay Area did much much
better than buying gold in 1970.',
 'created_utc': 1566311377,
 'gildings': {},
 'id': 'exhpyj3',
 'is_submitter': False,
 'link_id': 't3_csylne',
 'locked': False,
 'no_follow': True,
 'parent_id': 't3_csylne',
 'permalink': '/r/investing/comments/csylne/what_were_the_best_investments_
 of_the_stagflation/exhpyj3/',
```

```
'retrieved_on': 1566311379,
'score': 1,
'send_replies': True,
'stickied': False,
'subreddit': 'investing',
'subreddit_id': 't5_2qhhq',
'total_awards_received': 0}
```

You will learn more about applying NLP algorithms in Chapter 4, but for now let's just use an algorithm as a service platform called Algorithmia where you can access a large variety of algorithms based on machine learning and AI on text analysis, image manipulation, and so on by simply sending your data over a POST call on their REST API.

This service provides 10K free credits to everyone who signs up, and an additional 5K credits per month. This should be more than sufficient for running the example in Listing 1-4, since it will consume no more than 2–3 credits per request. Using more than the allotted free credits will incur a charge based on the request amount.

Once you register with Algorithmia, please go to the API keys section in the user dashboard and generate new API keys which you will use in this example.

Usually, you need to do some text preprocessing such as getting rid of new lines, special characters, and so on to get accurate text sentiments; but in this case, let's just take the text body and package it into a JSON format required by the sentiment analysis API (https://algorithmia.com/algorithms/specrom/GetSentimentsScorefromText).

The response is an id and a sentiment value from 0 to 1 where 0 and 1 mean very negative and positive sentiments, respectively. A value near to 0.5 indicates a neutral sentiment.

Listing 1-4. Creating request JSON

```
date_list = []
comment_list = []
rows_list = []
for i in range(len(json_dict["data"])):
    temp_dict = {}
    temp_dict["id"] = i
    temp_dict["text"] = json_dict["data"][i]['body']
    rows_list.append(temp_dict)
    date_list.append(json_dict["data"][i]['created_utc'])
```

```
    comment_list.append(json_dict["data"][i]['body'])
sample_dict = {}
sample_dict["documents"] = rows_list
payload = json.dumps(sample_dict)
with open("sentiments_payload.json", "w") as outfile:
    outfile.write(payload)
```

Creating an HTTP POST request needs a header parameter that sends over the authorization key and content type and a payload, which is a dictionary converted to JSON (see Listing 1-5).

Listing 1-5. Making a POST request

```
url = 'https://api.algorithmia.com/v1/algo/specrom/GetSentimentsScorefromText/
0.2.0?timeout=300'
headers = {

    'Authorization': YOUR_ALGORITHMIA_KEY,
    'content-type': "application/json",
    'accept': "application/json"
    }
response = requests.request("POST", url, data=payload, headers=headers)

print("Status Code: ", r.status_code)
print("*"*20)
print(r.headers)
# Output:
Status Code:  200
********************

{'Content-Encoding': 'gzip', 'Content-Type': 'application/json;
charset=utf-8', 'Date': 'Mon, 13 Apr 2020 11:08:58 GMT', 'Strict-Transport-
Security': 'max-age=86400; includeSubDomains', 'Vary': 'Accept-Encoding',
'X-Content-Type-Options': 'nosniff', 'X-Frame-Options': 'DENY', 'Content-
Length': '682', 'Connection': 'keep-alive'}
```

Let us load the response in a pandas dataframe and look at the first row to get an idea of the output (see Listing 1-6).

Listing 1-6. Viewing sentiments data

```python
import numpy as np
import pandas as pd

df_sent = pd.DataFrame(json.loads(response.text)["result"]["documents"])
df_sent.head(1)
#Output
```

	id	sentiments_score
0	0	0.523785

We should convert this score into distinct labels positive, negative, and neutral (see Listing 1-7).

Listing 1-7. Converting the sentiments score to labels

```python
def get_sentiments(score):
    if score > 0.6:
        return 'positive'
    elif score < 0.4:
        return 'negative'
    else:
        return 'neutral'

df_sent["sentiments"]=df_sent["sentiments_score"].apply(get_sentiments)
df_sent.head(1)
```

```
#Output
```

	id	sentiments_score	sentiments
0	0	0.523785	neutral

Finally, let us visualize the sentiments by plotting a bar plot as shown in Listing 1-8 and then displayed in Figure 1-10.

Listing 1-8. Plotting sentiments as a bar plot

```
import matplotlib.pyplot as plt
import seaborn as sns

sns.set()
%matplotlib inline

g = sns.countplot(df_sent["sentiments"])
loc, labels = plt.xticks()
g.set_xticklabels(labels, rotation=90)
g.set_title('Subreddit comments sentiment analysis')

g.set_ylabel("Count")
g.set_xlabel("Sentiments")
```

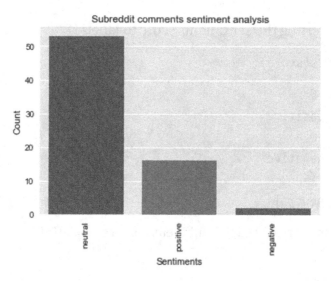

Figure 1-10. *Bar plot of sentiment analysis on subreddit comments*

So it seems like the comments are overwhelmingly neutral, with some positive comments and only a couple of negative comments.

Let us switch gears and see if these sentiments have any correlation with Exxon stock prices. We will get that using a REST API from www.alphavantage.co; it is free to use, but you will have to register to get a key (see Listing 1-9) from the alphavantage user dashboard.

Listing 1-9. Requesting data from the Alpha Vantage API

```
# Code block 1.2
# getting data from alphavantage

import requests
import json

test_url = 'https://www.alphavantage.co/query?function=TIME_SERIES_DAILY_AD
JUSTED&symbol=XOM&outputsize=full&apikey=' + API_KEY + '&datatype=csv'

r = requests.get(url = test_url)
print("Status Code: ", r.status_code)
print("*"*20)
print(r.headers)
html_response = r.text
with open("exxon_stock.csv", "w") as outfile:
    outfile.write(html_response)
# Output
Status Code:   200
********************
{'Connection': 'keep-alive', 'Server': 'gunicorn/19.7.0', 'Date': 'Thu, 16
Apr 2020 04:25:18 GMT', 'Transfer-Encoding': 'chunked', 'Vary': 'Cookie',
'X-Frame-Options': 'SAMEORIGIN', 'Allow': 'GET, HEAD, OPTIONS', 'Content-
Type': 'application/x-download', 'Content-Disposition': 'attachment;
filename=daily_adjusted_XOM.csv', 'Via': '1.1 vegur'}
```

This includes all the available stock prices data going back at least 10 years; hence, we will filter it to the date range we used for the previous sentiments (see Listing 1-10).

Listing 1-10. Parsing response data

```
import numpy as np
import pandas as pd

import matplotlib.pyplot as plt
import seaborn as sns

from dateutil import parser
```

```
datetime_obj = lambda x: parser.parse(x)

df = pd.read_csv("exxon_stock.csv", parse_dates=['timestamp'], date_
parser=datetime_obj)
start_date = pd.to_datetime(date_list[0], unit='s')
end_date = pd.to_datetime(date_list[-1], unit='s')
df = df[(df["timestamp"] >= start_date) & (df["timestamp"] <= end_date)]

df.head(1)
# Output
```

	timestamp	open	high	low	close	adjusted_ close	volume	dividend_ amount	split_ coefficient
86	2019-12-10	69.66	70.15	68.7	69.06	68.0723	14281286	0.0	1.0

As a final step, let's plot the closing price and volumes and see if the stock price stays neutral or not, as shown in Listing 1-11.

Listing 1-11. Plotting response data

```
# Plotting stock and volume

top = plt.subplot2grid((4,4), (0, 0), rowspan=3, colspan=4)
top.plot(df['timestamp'], df['close'], label = 'Closing price')
plt.title('Exxon Close Price')
plt.legend(loc=2)
bottom = plt.subplot2grid((4,4), (3,0), rowspan=1, colspan=4)
bottom.bar(df["timestamp"], df["volume"])
plt.title('Exxon Daily Trading Volume')
plt.gcf().set_size_inches(12,8)
plt.subplots_adjust(hspace=0.75)
```

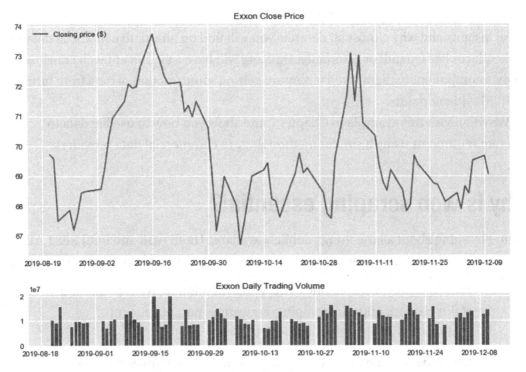

Figure 1-11. *Exxon stock prices and stock volumes*

As you can see from the plot shown in Figure 1-11, the stock prices have shown considerable movement in that five-month range with trading volume magnitudes higher than the number of comments extracted from a subreddit. So we can safely say that sentiment analysis of comments in just one subreddit is not a good indicator of the share price movement without performing any further trends analysis.

But that is hardly much of a surprise, since sentiment analysis only really works as a predictor if we are aggregating information from a large fraction of the visible Internet and plotting the data temporally as a time series to overlay it over stock market data.

There are lots of other flaws with simply plotting sentiments data like done earlier without correcting for the company-specific or sector-specific biases from the authors, editors, and so on. For example, someone who is a known environmentalist might have a well-known bias against fossil fuel companies like Exxon, and any negative sentiments expressed by such an author has to be corrected for that bias before using them as a predictor for stock market analysis.

This is a perfect illustration why we need to crawl on big data scale to generate useful insights and why almost all datasets you will find on alternative financial dataset marketplaces like Quandl or AlternativeData.org will have a significant web crawling and big data component to them, even if they are getting some fraction of data from hitting the REST API endpoints.

We will revisit this example in Chapter 7 and show you how to use big data to generate sentiments using a similar methodology to commercial data providers.

Why is web scraping essential?

So after learning about all the things publicly available (both paid and free) REST APIs can do for you, let me distill them into common use cases for performing web scraping:

- Your company works in one of the areas mentioned in the beginning of this chapter, and web scraping/crawling is part of your core business activity.

- The website you want to extract data from does not provide a public API, and there are no comparable third-party APIs which provide the same set of data you need.

- If there is an API, then the free tier is rate limited, meaning you are capped to calling it only a certain number of times. The paid tier of the API is cost prohibitive for your intended use case, but accessing the website itself is free.

- The API does not expose all the data you wish to obtain even in their paid tier, whereas the website contains that information.

How to turn web scraping into full-fledged product

Let us break down web scraping into its individual components:

- The first step is data ingestion, where all you are doing is grabbing the raw web pages from the Internet and storing them for further processing. I would argue that this is the easiest step in web crawling. We will perform web scraping and crawling using common Python-based parsing libraries in Chapter 2. We will also introduce cloud computing in Chapter 3 so that you are not restricted by memory and computational resources of your local server. We will discuss advanced crawling strategies in Chapter 8 which will bring together everything we have learned in the book.

- The second step is data processing, where we take in the raw data from web crawls and use some algorithms to extract useful information from it. In some cases, the algorithm will be as simple as traversing the HTML tree and extracting values of some tags such as the title and headline. In intermediate cases, we might have to run some pattern matching in addition to HTML parsing. For the most complicated use cases, we will have to run a gamut of NLP algorithms on raw text to extract people's names, contact details, text summaries, and so on. We will introduce natural language processing algorithms in Chapter 4, and we will put them into action in Chapters 6 and 7 on a Common Crawl dataset.

- The next step is loading the cleaned data from the preceding step into an appropriate database. For example, if your eventual products benefit from graph-based querying, then it's logical that you will load up the cleaned data onto a graph database such as Neo4j. On the other hand, if your product relies on providing full text searching, then it's logical to use a full text search database such as Elasticsearch or Apache Solr. For the majority of other uses, a general-purpose SQL database such as MySQL and PostgreSQL works well. We will introduce databases in Chapter 5 and illustrate practical applications in Chapters 6 and 7.

- The final step is exposing your database as a user client (mobile app and website) or allowing programmatic access through REST APIs. We will not talk about it; however, you can do it using the Amazon API Gateway.

Summary

We have introduced web scraping in this chapter and talked about quite a few real-world applications for it. We also discussed how to get structured data from third-party REST APIs using a Python-based library called requests.

CHAPTER 2

Web Scraping in Python Using Beautiful Soup Library

In this chapter, we'll go through the basic building blocks of web pages such as HTML and CSS and demonstrate scraping structured information from them using popular Python libraries such as Beautiful Soup and lxml. Later, we'll expand our knowledge and tackle issues that will make our scraper into a full-featured web crawler capable of fetching information from multiple web pages.

You will also learn about JavaScript and how it is used to insert dynamic content in modern web pages, and we will use Selenium to scrape information from JavaScript.

As a final piece, we'll take everything we have learned and use it to scrape information from the US FDA's warning letters database.

What are web pages all about?

All web pages are composed of HTML, which basically consists of plain text wrapped around tags that let web browsers know how to render the text. Examples of these tags include the following:

- Every HTML document starts and ends with <html>...</html> tags.

- By convention, <!DOCTYPE html> at the start of an HTML document. Note that any text wrapped in "<!" and ">" is considered to be a comment and not really rendered by web browsers.

- <head>...</head> encloses meta-information about the document.

© Jay M. Patel 2020
J. M. Patel, *Getting Structured Data from the Internet*, https://doi.org/10.1007/978-1-4842-6576-5_2

- <body>...</body> encloses the body of the document.

- <title>...</title> element specifies the title of the document.

- <h1>...</h1> to <h6>...</h6> tags are used for headers.

- <div>...</div> to indicate a division in an HTML document, generally used to group a set of elements.

- <p>...</p> to enclose a paragraph.

-
 to set a line break.

- <table>...</table> to start a table block.

  - <tr>...<tr/> is used for the rows.

  - <td>...</td> is used for individual cells.

- <img> for images.

- <a>...</a> for hyperlinks.

- <ul>...</ul>, <ol>...</ol> for unordered and ordered lists, respectively; inside of these, <li>...</li> is used for each list item.

HTML tags also contain common attributes enclosed within these tags:

- href attribute defines a hyperlink and anchor text and is enclosed by <a> tags.

  <a href="https://www.jaympatel.com">Jay M. Patel's homepage</a>

- Filename and location of images are specified by src attribute of the image tag.

  <img src="https://www.jaympatel.com/book_cover.jpg">

- It is very common to include width, height, and alternative text attributes in img tags for cases when the image cannot be displayed. You can also include a title attribute.

  <img src="https://www.jaympatel.com/book_cover.jpg" width="500" height="600" alt = "Jay's new web crawling book's cover image" title = "Jay' book cover">

- <html> tags also include a lang attribute.

  <html lang="en-US">

- A style attribute can also be included to specify a particular font color, size, and so on.

  ...

In addition to the HTML tags mentioned earlier, you can also optionally specify "ids" and "class" such as for h1 headers such as for h1 tags, such as

```
<h1 id="firstHeading" class="firstHeading" lang="en">Static sites are
awesome</h1>
```

- Id: A unique identifier representing a tag within the document
- Class: An identifier that can annotate multiple elements in a document and represents a space-separated series of Cascading Style Sheets (CSS) class names

Classes and ids are case sensitive, start with letters, and can include alphanumeric characters, hyphens, and underscores. A class may apply to any number of instances of any elements, whereas ID may only be applied to a single element within a document.

Classes and IDs are incredibly useful not only for applying styling via Cascading Style Sheets (CSS) (discussed in the next section) or using JavaScript but also for scraping useful information out of a page.

Let us create an HTML file: open your favorite text editor, copy-paste the code in Listing 2-1, and save it with a .html extension. I really like Notebook++ and it's free to download, but you can pretty much use any text editor you like.

Listing 2-1. Sample HTML code

```
<!DOCTYPE html>
<html>
<body>

<h1 id="firstHeading" class="firstHeading" lang="en">Getting Structured
Data from the Internet:</h1>
```

```
<h2>Running Web Crawlers/Scrapers on a Big Data Production Scale</h2>

<p id = "first">
Jay M. Patel
</p>

</body>

</html>
```

Once you have saved the file, simply double-click it, and it should open up in your browser. If you use Chrome or other major browsers like Firefox or Safari, right-click anywhere and select inspect, and then you will get the screen shown in Figure 2-1, which shows the source code you typed along with the rendered web page.

Figure 2-1. *Inspecting rendered HTML in Google Chrome*

Congratulations on creating your first HTML page! Let's insert some styling to the page.

Styling with Cascading Style Sheets (CSS)

Cascading Style Sheets (CSS) is a style sheet language used for describing the presentation of a document, such as layout, colors, and fonts written in a markup language like HTML. There are three ways to apply CSS styles to HTML pages:

- The first is inside a regular HTML tag such as shown next. You can also apply styles to change font colors: `<p style="color:green;">...</p>`. Using this type of styling will only affect the text enclosed by these tags. Note that inline styling takes precedence over other methods, and this is used sometimes to override the main CSS of the page.

```
<!DOCTYPE html>
<html>
<head>
<link rel="stylesheet" type="text/css" href="main.css">
</head>
<body>
```

- You can create a separate CSS file and link it by including it in a `link` tag within the main `<head>` of the HTML document; the browser will go out and request the CSS file whenever a page is loaded.

- Style can also be applied inside of `<style>...</style>` tags, placed inside the `<head>` tag of a page.

- A CSS file consists of code blocks which applies styling to individual HTML tags; in the following example, we are applying green color and center alignment to all text enclosed in the <p> paragraph tag:

```
p {
  color: green;
  text-align: center;
}
```

- We can use ID as a selector so that the styling is only applied to an id called 1para:

```
# 1para {
  color: green;
  text-align: center;
}
```

- You can also use a class to apply the same styling across all classes with value maincontent:

```
.maincontent {
  color: green;
  text-align: center;
}
```

- Let's combine two approaches for greater selectivity and apply style to only paragraphs within the maincontent's class:

```
p.maincontent {
  color: green;
  text-align: center;
}
```

Let us edit the preceding HTML file to add style="color:green;" to the <h1> tag. The revised HTML file with styling block is shown in Figure 2-2.

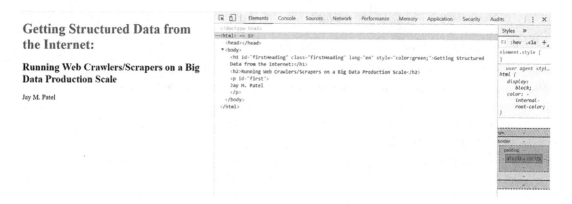

Figure 2-2. *Inspecting the HTML page with inline styling*

Scraping a web page with Beautiful Soup

Beautiful Soup is a Python library primarily intended to parse and extract information from an HTML string. It comes with a variety of HTML parsers that let us extract information even from a badly formatted HTML, which is unfortunately more common than what one assumes. We can use the requests library we already saw in Chapter 1 to fetch the HTML page, and once we have it in our local computer, we can start playing around with Beautiful Soup objects to extract useful information. As an initial example, let's simply scrape information from a Wikipedia page for (you guessed it) web scraping!

Web pages change all the time, and that makes it tricky when we are trying to learn web scraping which needs the web page to stay exactly the same as it was when I wrote this book so that even two or three years from now you can learn from live examples.

This is why web scraping book authors tend to host a small test website that can be used for scraping examples. I don't particularly like that approach since toy examples don't scale very well to real-world web pages, which are full of ill-formed HTML, unclosed tags, and so on. Besides, in a few years' time, maybe the author will stop hosting the pages on their website, and then how will readers work the examples in that case?

Therefore, ideally, we need to scrape from snapshots of real web pages with versioning so that a link will unambiguously refer to how the web page was on a particular date and time. Fortunately, such a resource already exists and is called the Internet Archive's Wayback Machine. We will be using links generated by the Wayback Machine so that you can continue to experiment and learn from this book even after 5–10 years since these links will stay up as long as the Internet Archive continues to exist.

It is easy enough to create a Beautiful Soup object, and in my experience, one of the easiest ways to find more information on a new object is to call the `dir()` on it to see all available methods and attributes.

As you can see, Beautiful Soup objects come with a long list of available methods with very intuitive names, such as FindParent, FindParents, findPreviousSibling, and findPreviousSiblings, for traversing the HTML tags, which presumably are helping you navigate the HTML tree (see Listing 2-2). There is no way for us to showcase all the methods here, but what we'll do is use a handful of them, and that will give you a sufficient idea on usage patterns for the rest of them.

Listing 2-2. Parsing HTML using the BeautifulSoup library

```
import requests
from bs4 import BeautifulSoup

test_url = 'https://web.archive.org/web/20200331040501/
https://en.wikipedia.org/wiki/Web_scraping'

r = requests.get(test_url)
html_response = r.text
# creating a beautifulsoup object
soup = BeautifulSoup(html_response,'html.parser')
print(type(soup))
print("*"*20)
print(dir(soup))
# output
<class 'bs4.BeautifulSoup'>
********************
['ASCII_SPACES', 'DEFAULT_BUILDER_FEATURES', 'HTML_FORMATTERS',
'NO_PARSER_SPECIFIED_WARNING', 'ROOT_TAG_NAME', 'XML_FORMATTERS',
'__bool__', '__call__', '__class__', '__contains__', '__copy__', '__
delattr__', '__delitem__', '__dict__', '__dir__', '__doc__', '__eq__',
'__format__', '__ge__', '__getattr__', '__getattribute__', '__getitem__',
'__getstate__', '__gt__', '__hash__', '__init__', '__init_subclass__',
'__iter__', '__le__', '__len__', '__lt__', '__module__', '__ne__', '__
new__', '__reduce__', '__reduce_ex__', '__repr__', '__setattr__', '__
setitem__', '__sizeof__', '__str__', '__subclasshook__', '__unicode__',
'__weakref__', '_all_strings', '_attr_value_as_string', '_attribute_
checker', '_check_markup_is_url', '_feed', '_find_all', '_find_one',
'_formatter_for_name', '_is_xml', '_lastRecursiveChild', '_last_
descendant', '_most_recent_element', '_popToTag', '_select_debug', '_
selector_combinators', '_should_pretty_print', '_tag_name_matches_and',
'append', 'attribselect_re', 'attrs', 'builder', 'can_be_empty_element',
'childGenerator', 'children', 'clear', 'contains_replacement_characters',
'contents', 'currentTag', 'current_data', 'declared_html_encoding',
'decode', 'decode_contents', 'decompose', 'descendants', 'encode', 'encode_
contents', 'endData', 'extract', 'fetchNextSiblings', 'fetchParents',
```

```
'fetchPrevious', 'fetchPreviousSiblings', 'find', 'findAll',
'findAllNext', 'findAllPrevious', 'findChild', 'findChildren', 'findNext',
'findNextSibling', 'findNextSiblings', 'findParent', 'findParents',
'findPrevious', 'findPreviousSibling', 'findPreviousSiblings',
'find_all', 'find_all_next', 'find_all_previous', 'find_next', 'find_
next_sibling', 'find_next_siblings', 'find_parent', 'find_parents',
'find_previous', 'find_previous_sibling', 'find_previous_siblings',
'format_string', 'get', 'getText', 'get_attribute_list', 'get_text',
'handle_data', 'handle_endtag', 'handle_starttag', 'has_attr', 'has_
key', 'hidden', 'index', 'insert', 'insert_after', 'insert_before',
'isSelfClosing', 'is_empty_element', 'is_xml', 'known_xml', 'markup',
'name', 'namespace', 'new_string', 'new_tag', 'next', 'nextGenerator',
'nextSibling', 'nextSiblingGenerator', 'next_element', 'next_elements',
'next_sibling', 'next_siblings', 'object_was_parsed', 'original_encoding',
'parent', 'parentGenerator', 'parents', 'parse_only', 'parserClass',
'parser_class', 'popTag', 'prefix', 'preserve_whitespace_tag_stack',
'preserve_whitespace_tags', 'prettify', 'previous', 'previousGenerator',
'previousSibling', 'previousSiblingGenerator', 'previous_element',
'previous_elements', 'previous_sibling', 'previous_siblings', 'pushTag',
'quoted_colon', 'recursiveChildGenerator', 'renderContents', 'replaceWith',
'replaceWithChildren', 'replace_with', 'replace_with_children', 'reset',
'select', 'select_one', 'setup', 'string', 'strings', 'stripped_strings',
'tagStack', 'tag_name_re', 'text', 'unwrap', 'wrap']
```

The second major object created by the Beautiful Soup library is known as a tag object, which corresponds to the HTML/XML tag in the original document. Let us call the tag object for h1 heading; a tag's name can be accessed by .name method, and the attributes can be accessed via treating it as a dictionary. So in the case shown in Listing 2-3, I can access the tag id by simply calling first_tag["id"]; to get all available attributes, please review the .attrs method.

Listing 2-3. Exploring BeautifulSoup objects

```
first_tag = (soup.h1)
print(type(first_tag))
print("*"*20)
print(first_tag)
print("*"*20)
print(first_tag["id"])
print("*"*20)
print(first_tag.attrs)

# Output

<class 'bs4.element.Tag'>
********************
<h1 class="firstHeading" id="firstHeading" lang="en">Web scraping</h1>
********************
firstHeading
********************
{'id': 'firstHeading', 'class': ['firstHeading'], 'lang': 'en'}
```

The last Beautiful Soup object of interest is the NavigableString type, and contains the string that is enclosed by HTML/XML tags. You can easily convert this to a regular Python string by calling the str() method on it as shown in Listing 2-4. An analogous way to get the Python string is by simply calling the get_text() method on the tag object, and this is actually the preferred way to do it; we went through this exercise just to make you familiar with all the objects of the Beautiful Soup library.

Listing 2-4. Exploring BeautifulSoup objects (cont.)

```
first_string = first_tag.string
print(type(first_string))
print("*"*20)
python_string = str(first_string)
print(type(python_string), python_string)
print("*"*20)
print(type(first_tag.get_text()), first_tag.get_text())
```

```
# Output

<class 'bs4.element.NavigableString'>
********************
<class 'str'> Web scraping
********************
<class 'str'> Web scraping
```

find() and find_all()

These are some of the most versatile methods in Beautiful Soup; find_all() retrieves matching tags from all the nested HTML tags (called descendants), and if you pass in a list, then it will retrieve all the matching objects. Let us use find_all() to get contents enclosed by the h1 and h2 tags from the wiki page, as shown in Listing 2-5.

In contrast, the find() method will only return the first matching instance and will ignore all the remaining arguments.

Listing 2-5. Exploring the find_all function

```
# Passing a list to find_all method
for object in soup.find_all(['h1', 'h2']):
    print(object.get_text())
# doing the same to find()
print("*"*20)
print(soup.find(['h1','h2']).get_text())

# Output:

Web scraping
Contents
History[edit]
Techniques[edit]
Software[edit]
Legal issues[edit]
Methods to prevent web scraping[edit]
See also[edit]
References[edit]
Navigation menu
********************

Web scraping
```

Getting links from a Wikipedia page

Let's say that you are trying to scrape the anchor text and links to the "see also" section of the preceding Wikipedia page (as shown in Figure 2-3).

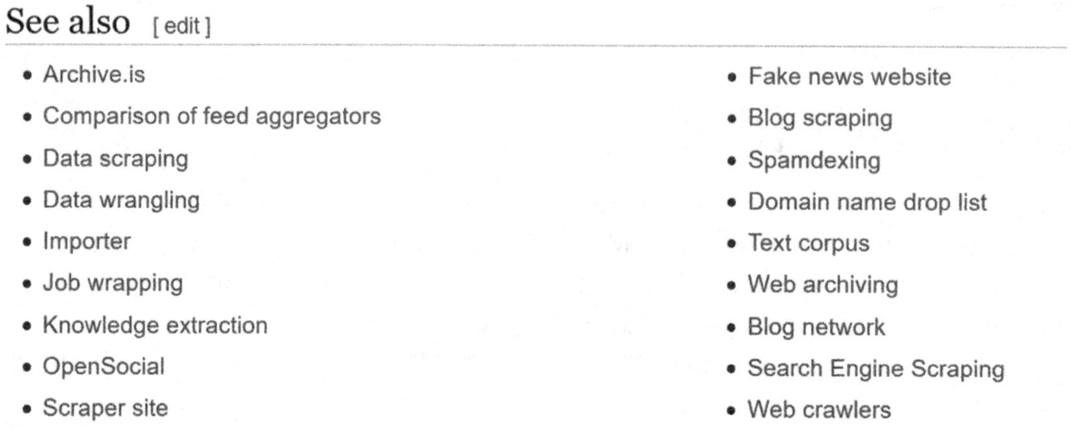

See also [edit]

- Archive.is
- Comparison of feed aggregators
- Data scraping
- Data wrangling
- Importer
- Job wrapping
- Knowledge extraction
- OpenSocial
- Scraper site

- Fake news website
- Blog scraping
- Spamdexing
- Domain name drop list
- Text corpus
- Web archiving
- Blog network
- Search Engine Scraping
- Web crawlers

Figure 2-3. *Screenshot of links and text you wish to scrape*

The first step would be to locate these links in the source code of the HTML page so as to find the class name or a CSS style, which can help you target this using Beautiful Soup's find and find_all() methods. We used the inspect in Chrome to find out that the div class we are interested in is "div-col columns column-width."

Listing 2-6. Extracting links

```
link_div = soup.find('div', {'class':'div-col columns column-width'})
link_dict = {}
links = link_div.find_all('a')

for link in links:
    anchor_text = link.get_text()
    link_dict[anchor_text] = link['href']
print(link_dict)
# output
```

```
{'Archive.is': '/wiki/Archive.is', 'Comparison of feed aggregators': '/
wiki/Comparison_of_feed_aggregators', 'Data scraping': '/wiki/Data_
scraping', 'Data wrangling': '/wiki/Data_wrangling', 'Importer': '/wiki/
Importer_(computing)', 'Job wrapping': '/wiki/Job_wrapping', 'Knowledge
extraction': '/wiki/Knowledge_extraction', 'OpenSocial': '/wiki/
OpenSocial', 'Scraper site': '/wiki/Scraper_site', 'Fake news website':
'/wiki/Fake_news_website', 'Blog scraping': '/wiki/Blog_scraping',
'Spamdexing': '/wiki/Spamdexing', 'Domain name drop list': '/wiki/Domain_
name_drop_list', 'Text corpus': '/wiki/Text_corpus', 'Web archiving':
'/wiki/Web_archiving', 'Blog network': '/wiki/Blog_network', 'Search
Engine Scraping': '/wiki/Search_Engine_Scraping', 'Web crawlers': '/wiki/
Category:Web_crawlers'}
```

The first line of the code in Listing 2-6 finds all the <div> tags with the class name as "div-col columns column-width"; the resulting object link_div is a Beautiful Soup <tag> object. Next, we are using this tag object and calling a find_all() to find all the instances with <a> HTML tag which encloses an anchor text and a link. Once we have a list of such Beautiful Soup tag objects, all we need to do is iterate through them to pull out the anchor text and the link which is accessible by the "hrefs" links. We are loading it onto a Python dictionary which you can easily save as JSON, thus extracting structured information from the scraped Wikipedia page. Note that the links extracted are relative links, but you can simply use Python string methods to append the baseUrl with each of the links to get an absolute URL.

Scrape an ecommerce store site

Extracting structured information from ecommerce websites for price and competitor monitoring is in fact one of the major use cases for web scraping.

You can view the headers your browser is sending as part of request headers by going over to a site such as www.whatismybrowser.com. My request header's user-agent is shown in the screenshot in Figure 2-4.

ACCEPT	text/html,application/xhtml+xml,application /xml;q=0.9,*/*;q=0.8
ACCEPT-ENCODING	gzip, deflate, br
ACCEPT-LANGUAGE	en-US,en;q=0.5
HOST	www.whatismybrowser.com
REFERER	https://www.google.com/
UPGRADE-INSECURE-REQUESTS	1
USER-AGENT	Mozilla/5.0 (Windows NT 10.0; Win64; x64; rv:52.0) Gecko/20100101 Firefox/52.0

Figure 2-4. *Browser headers*

I would encourage you to modify your requests from now on and include a header dictionary which includes a user-agent so that you can blend in with real humans using browsers when you are programmatically accessing the sites for web scraping. There are much more advanced antiscraping measures websites can take, so this will not fool everyone, but this will get you more access than having no headers at all. To illustrate an effective antiscraping measure, let us try to scrape from Amazon.com; in Listing 2-7, all we are doing is removing scripts from the BeautifulSoup object and converting the soup object into full text. As you can see, Amazon correctly identified that we are a robot and gave us a CAPTCHA instead of allowing us to proceed with the page.

Listing 2-7. Scraping from Amazon.com

```
my_headers = {
'User-Agent': 'Mozilla/5.0 (Windows NT 10.0; Win64; x64) AppleWebKit/537.36
' + ' (KHTML, like Gecko) Chrome/61.0.3163.100Safari/537.36'
}

url = 'https://www.amazon.com'
rr = requests.get(url, headers = my_headers)
ht_response = rr.text
soup = BeautifulSoup(ht_response,'html.parser')
for script in soup(["script"]):
        script.extract()
soup.get_text()
```

```
# Output
"\n\n\n\n\n\n\n\n\nRobot Check\n\n\n\n\n\n\n\n\n\n\n\n\n\nEnter the
characters you see below\nSorry, we just need to make sure you're not
a robot. For best results, please make sure your browser is accepting
cookies.\n\n\n\n\n\n\n\n\n\nType the characters you see in this
image:\n\n\n\n\n\n\n\n\nTry different image\n\n\n\n\n\n\n\n\n\n\
nContinue shopping\n\n\n\n\n\n\n\n\n\n\nConditions of Use\n\n\n\
nPrivacy Policy\n\n\n          © 1996-2014, Amazon.com, Inc. or its
affiliates\n          \n\n\n\n\n\n\n\n"
```

Let us switch gears and instead try to extract all the links visible on the first page of the ecommerce site of Apress. We will be using an Internet Archive snapshot (Listing 2-8). We are only filtering links to extract only from class name product information so that our links correspond to individual book pages.

Listing 2-8. Scraping from the Apress ecommerce store

```
url = 'https://web.archive.org/web/20200219120507/https://www.apress.com/
us/shop'
base_url = 'https://web.archive.org/web/20200219120507'
my_headers = {
'User-Agent': 'Mozilla/5.0 (Windows NT 10.0; Win64; x64) AppleWebKit/537.36
' + ' (KHTML, like Gecko) Chrome/61.0.3163.100Safari/537.36'
}
r = requests.get(url, headers = my_headers)
ht_response = r.text

soup = BeautifulSoup(ht_response,'html.parser')

product_info = soup.find_all("div", {"class":"product-information"})
url_list =[]

for product in product_info:
    temp_url = base_url + str(product.parent.find('a')["href"])
    url_list.append(temp_url)
```

Let's take one URL from this list and extract the book name, book format, and price from it (Listing 2-9).

Listing 2-9. Extracting structured information from a URL

```
url = 'https://web.archive.org/web/20191018112156/https://www.apress.com/
us/book/9781484249406'

my_headers = {
'User-Agent': 'Mozilla/5.0 (Windows NT 10.0; Win64; x64) AppleWebKit/537.36
' + ' (KHTML, like Gecko) Chrome/61.0.3163.100Safari/537.36'
}

rr = requests.get(url, headers = my_headers)
ht_response = rr.text

temp_dict = {}
results_list = []
main_dict = {}

soup = BeautifulSoup(ht_response,'html.parser')
primary_buy = soup.find("span", {"class":"cover-type"})
temp_dict["book_type"] = primary_buy.get_text()
temp_dict["book_price"] = primary_buy.parent.find("span", {"class":
"price"}).get_text().strip()
temp_dict["book_name"] = soup.find('h1').get_text()
temp_dict["url"] = url

results_list.append(temp_dict)
main_dict["extracted_products"] = results_list

print(main_dict)
# Output
{'extracted_products': [{'book_type': 'eBook', 'book_price': '$39.99',
'book_name': 'Pro .NET Benchmarking', 'url': 'https://web.archive.org/
web/20191018112156/https://www.apress.com/us/book/9781484249406'}]}
```

Profiling Beautiful Soup parsers

We have refrained from talking about performance in the previous section since we mainly wanted you to first get an idea of the capabilities of the Beautiful Soup library.

If you look at Listing 2-9, you will immediately see that there is very little we can do about how long it takes to fetch the HTML page using the requests library since that is totally in the hands of how much bandwidth we have and the server's response time.

So the only other thing we can profile is the Beautiful Soup library itself. It's a powerful way to access almost any object in HTML, and it definitely has its place in the web scraping toolbox.

However, it's the slow HTML parsing speed that makes it unviable for large-scale web crawling loads.

You can get some performance boost by switching to lxml parser, but it still isn't much compared to parsing the DOM using XPath as discussed in the next section.

Let's use Python's built-in profiler (cProfile) to identify the most time-consuming function calls using the default html.parser (Listing 2-10).

Listing 2-10. Profiling Beautiful Soup parsers

```
import cProfile

cProfile.run('''
temp_dict = {}
results_list = []
main_dict = {}
def main():
        soup = BeautifulSoup(ht_response,'html.parser')
        primary_buy = soup.find("span", {"class":"cover-type"})
        temp_dict["book_type"] = primary_buy.get_text()
        temp_dict["book_price"] = primary_buy.parent.find("span", {"class":
        "price"}).get_text().strip()
        temp_dict["book_name"] = soup.find('h1').get_text()
        temp_dict["url"] = url
        results_list.append(temp_dict)
```

47

```
        main_dict["extracted_products"] = results_list
        return(results_list)
main()''', 'restats')

#https://docs.python.org/3.6/library/profile.html
import pstats
p = pstats.Stats('restats')
p.sort_stats('cumtime').print_stats(15)

#Output
```

Sun Apr 19 09:39:00 2020   restats

         79174 function calls (79158 primitive calls) in 0.086 seconds

   Ordered by: cumulative time
   List reduced from 102 to 15 due to restriction <15>

ncalls	tottime	percall	cumtime	percall	filename:lineno(function)
1	0.000	0.000	0.086	0.086	{built-in method builtins.exec}
1	0.000	0.000	0.085	0.085	\<string>:2(\<module>)
1	0.000	0.000	0.085	0.085	\<string>:5(main)
1	0.000	0.000	0.078	0.078	C:\ProgramData\Anaconda3\lib\ site-packages\bs4\ __init__.py:87(__init__)
1	0.000	0.000	0.078	0.078	C:\ProgramData\Anaconda3\lib\ site-packages\bs4\ __init__.py:285(_feed)
1	0.000	0.000	0.078	0.078	C:\ProgramData\Anaconda3\lib\ site-packages\bs4\builder\ _htmlparser.py:210(feed)
1	0.000	0.000	0.078	0.078	C:\ProgramData\Anaconda3\lib\ html\parser.py:104(feed)
1	0.007	0.007	0.078	0.078	C:\ProgramData\Anaconda3\lib\ html\parser.py:134(goahead)
715	0.008	0.000	0.045	0.000	C:\ProgramData\Anaconda3\ lib\html\parser.py:301(parse_ starttag)

715	0.003	0.000	0.027	0.000	C:\ProgramData\Anaconda3\lib\ site-packages\bs4\builder\ _htmlparser.py:79(handle _starttag)
715	0.002	0.000	0.023	0.000	C:\ProgramData\Anaconda3\lib\ site-packages\bs4__init __.py:447(handle_starttag)
619	0.003	0.000	0.015	0.000	C:\ProgramData\Anaconda3\lib\ html\parser.py:386(parse_endtag)
1464	0.005	0.000	0.014	0.000	C:\ProgramData\Anaconda3\lib\ site-packages\bs4__init __.py:337(endData)
714	0.001	0.000	0.012	0.000	C:\ProgramData\Anaconda3\lib\ site-packages\bs4\builder\ _htmlparser.py:107(handle_endtag)
716	0.003	0.000	0.011	0.000	C:\ProgramData\Anaconda3\lib\ site-packages\bs4\element.py: 813(__init__)

This should print out the output consisting of the top 15 most time-consuming calls. Now, there are calls going to bs4__init__.py that we won't be able to optimize without a major refactoring of the library; the next top time-consuming calls are all made by html\parser.py.

Let us profile the main function again with the only modification that we have switched out the parser to lxml. I am only showing the output in Listing 2-11.

Listing 2-11. Profiling Beautiful Soup parsers (cont.)

```
# Output:
Sun Apr 19 09:39:57 2020    restats

        63900 function calls (63880 primitive calls) in 0.064 seconds

Ordered by: cumulative time
List reduced from 168 to 15 due to restriction <15>
```

ncalls	tottime	percall	cumtime	percall	filename:lineno(function)
1	0.000	0.000	0.064	0.064	{built-in method builtins.exec}
1	0.000	0.000	0.063	0.063	<string>:2(<module>)
1	0.000	0.000	0.063	0.063	<string>:5(main)
1	0.000	0.000	0.058	0.058	C:\ProgramData\Anaconda3\lib\site-packages\bs4__init__.py:87(__init__)
1	0.000	0.000	0.058	0.058	C:\ProgramData\Anaconda3\lib\site-packages\bs4__init__.py:285(_feed)
1	0.000	0.000	0.058	0.058	C:\ProgramData\Anaconda3\lib\site-packages\bs4\builder_lxml.py:246(feed)
2/1	0.006	0.003	0.047	0.047	src/lxml/parser.pxi:1242(feed)
715	0.001	0.000	0.026	0.000	src/lxml/saxparser.pxi:374(_handleSaxTargetStartNoNs)
715	0.000	0.000	0.024	0.000	src/lxml/saxparser.pxi:401(_callTargetSaxStart)
715	0.000	0.000	0.024	0.000	src/lxml/parsertarget.pxi:78(_handleSaxStart)
715	0.004	0.000	0.023	0.000	C:\ProgramData\Anaconda3\lib\site-packages\bs4\builder_lxml.py:145(start)
715	0.002	0.000	0.017	0.000	C:\ProgramData\Anaconda3\lib\site-packages\bs4__init__.py:447(handle_starttag)
715	0.001	0.000	0.011	0.000	src/lxml/saxparser.pxi:452(_handleSaxEndNoNs)
2181	0.004	0.000	0.011	0.000	C:\ProgramData\Anaconda3\lib\site-packages\bs4__init__.py:337(endData)
715	0.000	0.000	0.010	0.000	src/lxml/parsertarget.pxi:84(_handleSaxEnd)

<pstats.Stats at 0x2a852202780>

You can clearly see a reduction in not only the number of function calls but also the cumulative time, and most of those time advantages are directly coming from using the lxml-based parser builder_lxml.py as the back end for Beautiful Soup.

XPath

XPath stems its origins in the XSLT standard and stands for XML path language. Its syntax allows you to identify paths and nodes of an XML (and HTML) document. You will almost never have to write your own XPath from scratch so we will not spend any time talking about the XPath syntax, but you are encouraged to go through the XPath 3.1 standard (www.w3.org/TR/xpath-31/) for complete details.

The most common way we find XPath is by taking the help of developer tools in Google Chrome. For example, if I want the XPath to be the price of a book on the Apress site, I will right-click anywhere on the page and click inspect. Once there, click the element you want the XPath for; in our case, we want the price of a particular book (see Figure 2-5). Now, you can click copy and either select the abbreviated XPath or the complete XPath of a particular object; you can use either of that for web scraping.

Abbreviated XPath: //*[@id="id2"]/div/div/div/ul/li[3]/div[2]/span[2]/span

Complete XPath: /html/body/div[5]/div/div/div/div/div[3]/div/div/div/ul/li[3]/div[2]/span[2]/span

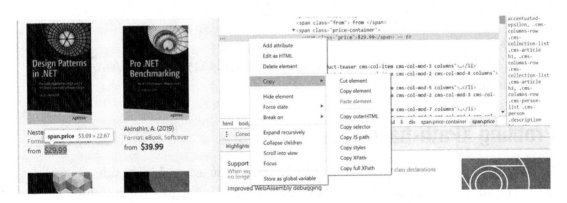

Figure 2-5. *XPath for the Apress ecommerce store*

We will use the XPath syntax to extract the same information as Listing 2-12.

Listing 2-12. Using the lxml library

```
from lxml.html import fromstring, tostring
temp_dict = {}
results_list = []
main_dict = {}
def main():

        tree = fromstring(ht_response)

        temp_dict["book_type"] = tree.xpath('//*[@id="content"]/div[2]/
        div[2]/div[1]/div/dl/dt[1]/span[1]/text()')[0]
        temp_dict["book_price"] = tree.xpath('//*[@id="content"]/div[2]/
        div[2]/div[1]/div/dl/dt[1]/span[2]/span/text()')[0].strip()
        temp_dict["book_name"] = tree.xpath('//*[@id="content"]/div[2]/
        div[1]/div[1]/div[1]/div[2]/h1/text()')[0]
        temp_dict["url"] = url
        results_list.append(temp_dict)

        main_dict["extracted_products"] = results_list
        return(main_dict)
main()
#Output
{'extracted_products': [{'book_name': 'Pro .NET Benchmarking',
    'book_price': '$39.99',
    'book_type': 'eBook',
    'url': 'https://web.archive.org/web/20191018112156/https://www.apress.
    com/us/book/9781484249406'}]}
```

Profiling XPath-based lxml

Profiling the main() function from Listing 2-12 gives us an astonishing result; we are getting fivefold time improvement and a drastic 160-fold reduction in the number of function calls.

Even if we end up parsing 100,000 documents of similar type, it will only take us 26.67 minutes (0.44 hrs) vs. 143.33 minutes (2.39 hrs) for Beautiful Soup.

I just wanted to put this out there so that you know that even though we are using Beautiful Soup here for examples, you should strongly consider switching to

XPath-based parsing once your workload gets into parsing hundreds of thousands of web pages (see Listing 2-13).

Listing 2-13. Profiling the lxml library

```
Sun Apr 19 10:08:05 2020     restats

        436 function calls in 0.016 seconds

Ordered by: cumulative time
List reduced from 103 to 15 due to restriction <15>

ncalls  tottime  percall  cumtime  percall filename:lineno(function)
     1    0.000    0.000    0.016    0.016 {built-in method builtins.exec}
     1    0.000    0.000    0.015    0.015 <string>:2(<module>)
     1    0.000    0.000    0.015    0.015 <string>:5(main)
     1    0.000    0.000    0.012    0.012 C:\ProgramData\Anaconda3\lib\
                                            site-packages\lxml\html\
                                            __init__.py:861(fromstring)
     1    0.000    0.000    0.012    0.012 C:\ProgramData\Anaconda3\
                                            lib\site-packages\lxml\
                                            html\__init__.py:759(document_
                                            fromstring)
     1    0.000    0.000    0.012    0.012 src/lxml/etree.
                                            pyx:3198(fromstring)
     1    0.007    0.007    0.007    0.007 src/lxml/etree.pyx:354(getroot)
     1    0.000    0.000    0.005    0.005 src/lxml/parser.pxi:1869
                                            (_parseMemoryDocument)
     1    0.000    0.000    0.005    0.005 src/lxml/parser.pxi:1731
                                            (_parseDoc)
     1    0.005    0.005    0.005    0.005 src/lxml/parser.pxi:1009
                                            (_parseUnicodeDoc)
     3    0.000    0.000    0.003    0.001 src/lxml/etree.pyx:1568(xpath)
     3    0.003    0.001    0.003    0.001 src/lxml/xpath.pxi:281(__call__)
     3    0.000    0.000    0.000    0.000 src/lxml/xpath.pxi:252(__init__)
     3    0.000    0.000    0.000    0.000 src/lxml/xpath.pxi:131(__init__)
    30    0.000    0.000    0.000    0.000 src/lxml/parser.pxi:612
                                            (_forwardParserError)
```

Crawling an entire site

We will discuss important parameters before we can start crawling entire websites. Let us start out by writing a naive crawler, point out its shortcomings, and try to fix it by specific solutions.

Essentially, we have one function called link_crawler() which takes in a seed_url, and it uses that to request the first page. Once the links are parsed, we can start loading them into the initial set of URLs to be crawled.

As we start getting down the list, we will see that there are pages we have already requested and parsed, and to keep track of those, we have another set called seen_url_list.

We are trying to restrict our crawl size, so that we are restricting domain addresses to only those which are from the seed list; another way we have restricted the crawl rate is by specifying a max_n number which refers to the number of pages we have fetched (see Listing 2-14). We are also taking care of relative links and adding a base URL.

Listing 2-14. Link crawler

```
import requests
from bs4 import BeautifulSoup

def link_crawler(seed_url, max_n = 5000):
    my_headers = {
    'User-Agent': 'Mozilla/5.0 (Windows NT 10.0; Win64; x64) AppleWebKit/
    537.36 ' + ' (KHTML, like Gecko) Chrome/61.0.3163.100Safari/537.36'
    }
    initial_url_set = set()
    initial_url_set.add(seed_url)
    seen_url_set = set()

    while len(initial_url_set)!=0 and len(seen_url_set) < max_n:
        temp_url = initial_url_set.pop()
        if temp_url in seen_url_set:
            continue
        else:
            seen_url_set.add(temp_url)
            r = requests.get(url = temp_url, headers = my_headers)
            st_code = r.status_code
```

```python
        html_response = r.text
        soup = BeautifulSoup(html_response,'html.parser')
        links = soup.find_all('a', href=True)
        for link in links:

            if ('http' in link['href']):
                if seed_url.split(".")[1] in link['href']:
                    initial_url_set.add(link['href'])

            elif [char for char in link['href']][0] == '/':
                final_url = seed_url+link['href']
                initial_url_set.add(final_url)

    return(initial_url_set, seen_url_set)
seed_url = 'http://www.jaympatel.com'
link_crawler(seed_url)
#output:
(set(),
 {'http://jaympatel.com/',
  'http://jaympatel.com/2018/11/get-started-with-git-and-github-in-under-
  10-minutes/',
  'http://jaympatel.com/2019/02/introduction-to-natural-language-
  processing-rule-based-methods-name-entity-recognition-ner-and-text-
  classification/',
  'http://jaympatel.com/2019/02/introduction-to-web-scraping-in-python-
  using-beautiful-soup/',
  'http://jaympatel.com/2019/02/natural-language-processing-nlp-term-
  frequency-inverse-document-frequency-tf-idf-based-vectorization-in-
  python/',
  'http://jaympatel.com/2019/02/natural-language-processing-nlp-text-
  vectorization-and-bag-of-words-approach/',
  'http://jaympatel.com/2019/02/natural-language-processing-nlp-word-
  embeddings-words2vec-glove-based-text-vectorization-in-python/',
  'http://jaympatel.com/2019/02/top-data-science-interview-questions-and-
  answers/',
  'http://jaympatel.com/2019/02/using-twitter-rest-apis-in-python-to-
  search-and-download-tweets-in-bulk/',
```

```
'http://jaympatel.com/2019/02/why-is-web-scraping-essential-and-who-uses-
web-scraping/',
'http://jaympatel.com/2020/01/introduction-to-machine-learning-metrics/',
'http://jaympatel.com/about/',
'http://jaympatel.com/books',
'http://jaympatel.com/books/',
'http://jaympatel.com/categories/',
'http://jaympatel.com/categories/#data-mining',
'http://jaympatel.com/categories/#data-science',
'http://jaympatel.com/categories/#interviews',
'http://jaympatel.com/categories/#machine-learning',
'http://jaympatel.com/categories/#natural-language-processing',
'http://jaympatel.com/categories/#requests',
'http://jaympatel.com/categories/#sentiments',
'http://jaympatel.com/categories/#software-development',
'http://jaympatel.com/categories/#text-vectorization',
'http://jaympatel.com/categories/#twitter',
'http://jaympatel.com/categories/#web-scraping',
'http://jaympatel.com/consulting-services',
'http://jaympatel.com/consulting-services/',
'http://jaympatel.com/cv',
'http://jaympatel.com/cv/',
'http://jaympatel.com/pages/CV.pdf',
'http://jaympatel.com/tags/',
'http://jaympatel.com/tags/#coefficient-of-determination-r2',
'http://jaympatel.com/tags/#git',
'http://jaympatel.com/tags/#glove',
'http://jaympatel.com/tags/#information-criterion',
'http://jaympatel.com/tags/#language-detection',
'http://jaympatel.com/tags/#machine-learning',
'http://jaympatel.com/tags/#name-entity-recognition',
'http://jaympatel.com/tags/#p-value',
'http://jaympatel.com/tags/#regex',
'http://jaympatel.com/tags/#regression',
'http://jaympatel.com/tags/#t-test',
'http://jaympatel.com/tags/#term-frequency-inverse-document-frequency-tf-idf',
```

```
'http://jaympatel.com/tags/#text-mining',
'http://jaympatel.com/tags/#tweepy',
'http://jaympatel.com/tags/#version-control',
'http://jaympatel.com/tags/#web-scraping',
'http://jaympatel.com/tags/#word-embeddings',
'http://jaympatel.com/tags/#words2vec',
'http://www.jaympatel.com/assets/DoD_SERDP_case_study.pdf'})
```

The function in Listing 2-14 works fine for testing and educational purposes, but it has some serious shortcomings which make it entirely unsuitable for using it regularly. Let us go through some of the issues and see how we can make it robust enough for practical uses.

URL normalization

In general, when we are setting up a crawler, we are only looking to scrape information from a specific type of pages. For example, we typically exclude scraping from links which point to CSS sheets or JavaScript. You can get a much more granular idea on the filetype at a particular link by checking the content-type in the request header, but this requires you to actually ping the link which is not practical in many cases.

Another common scenario is normalizing multiple links which all are in fact pointing to one page. These days, single-page HTML sites are becoming very common, where a user can jump through different sections of the page using anchor links. For example, all the following links are pointing to different sections of the same page:

```
<a href="#pricing">Pricing</a><br />
```

```
<a href="#license-cost">License Cost</a></li>
```

Another way the same link may get different URLs is through Urchin Tracking Module (UTM) parameters which are commonly used for tracking campaigns in digital marketing and are pretty common on the Web. As an example, let us consider the following two URLs for Specrom Analytics with UTM parameters, with the only difference being the utm_source parameter:

```
www.specrom.com/?utm_source=newsletter&utm_medium=banner&utm_
campaign=fall_sale&utm_term=web%20scraping%20crawling
```

```
www.specrom.com/?utm_source=google&utm_medium=banner&utm_campaign=fall_
sale&utm_term=web%20scraping%20crawling
```

Both links are pointing to www.specrom.com (you can verify it if you want); so if your crawler took in the URLs, then you will end up with three copies of the same page which will waste your bandwidth and computing not only to fetch them but also down the road when you try to deduplicate your database.

There is also a question of trailing slashes; traditionally, web addresses with trailing slashes indicated folders, whereas the ones without it indicated files. This definitely doesn't hold true anymore, but we are still stuck with pages with and without slashes both pointing to the same content. Google has issued a guidance for webmasters about this issue, and their preferred way is a 301 redirect from a duplicate page to the canonical one. To keep things simple, we will simply ignore trailing slashes in our code.

Therefore, you will need to incorporate URL normalization in your link crawler; in our case, we can simply exclude everything after #-[?*!@=]. You can easily accomplish this by using regular expressions or by using Python's string methods; but in our case, we will use the Python package tld which has a handy attribute called parsed URL to get rid of fragments and queries from the URL (Listing 2-15).

Listing 2-15. URL normalization

```
from tld import get_tld
sample_url = 'http://www.specrom.com/?utm_source=google&utm_
medium=banner&utm_campaign=fall_sale&utm_term=web%20scraping%20crawling'

def get_normalized_url(url):
    res = get_tld(url, as_object=True)
    path_list = [char for char in res.parsed_url.path]
    if len(path_list) == 0:
        final_url = res.parsed_url.scheme+'://'+res.parsed_url.netloc
        elif path_list[-1] == '/':
        final_string = ''.join(path_list[:-1])
        final_url = res.parsed_url.scheme+'://'+res.parsed_url.
        netloc+final_string
    else:

        final_url = url
    return final_url

get_normalized_url(sample_url)
#output
'http://www.specrom.com'
```

Robots.txt and crawl delay

We can use our URL link finder function to crawl the entire website; but first we will have to make some modifications to ensure that we are not overstepping the scope of legitimate crawling.

Most sitemasters put a file called robots.txt in the path `http://www.example.com/robots.txtm` which explicitly lists out directories and pages on their site which are OK to crawl and what parts are off limits to crawlers. These are just a suggestion, and you can scrape a website that explicitly prohibits crawling using the robots.txt file, but it's unethical and against terms of use that can open you up for a legal challenge in some jurisdictions. Some robots.txt files also try to help crawlers by including a sitemap so that you can build a URL index from it.

If a particular site is very open to crawling, then it will simply put

```
User-agent: *
Disallow:
```

On the other end of the spectrum, if a site doesn't want to be crawled by anyone, including search engines like Google and Bing, then it can put this:

```
User-agent: *
Disallow: /
```

Between these two extreme cases, we find that most sites are open to crawling their sites except perhaps some pages such as login screens and other private pages.

You will see some websites explicitly single out specific crawlers and restrict their sites from them compared to other crawlers. One common example is that of robots.txt file at Amazon.com (`www.amazon.com/robots.txt`) which mentions the following:

```
User-agent: EtaoSpider
Disallow: /
```

eTao is an ecommerce product search engine by Taobao (owned by Alibaba Group), which is one of the biggest search engines in China. Amazon has blocked its site completely to eTaoSpider presumably because Amazon does not want its data to be used by eTao for price comparisons.

Another important parameter of the robots.txt file is called crawl delay. This is used to specify time intervals in seconds between fetching successive pages. Googlebot doesn't support this, but you can set the crawl delay through Google Search Console. Bing and other search engines still support the crawl delay parameter. An example of this can be found at `https://camelcamelcamel.com/robots.txt` shown next:

```
User-Agent: bingbot
  Crawl-delay: 2
```

In other words, a delay of 2 seconds per fetched page translates into a maximum of 43,200 pages per day.

I cannot emphasize how important it is to set a reasonable time between fetches even when no crawl delay is explicitly specified through robots.txt. Almost all crawlers discussed in this book can easily bring down a small website if you don't set a crawl delay parameter. In the black hat world, taking down of websites by flooding with indiscriminate traffic is called a distributed denial-of-service (DDoS) attack, and your crawler will indeed by doing a DDoS attack even if you didn't intend it if you don't explicitly restrain it by setting limits of the number of pages fetched per second. This is especially true once you launch your crawler parallelized fashion with a distributed framework using rotating proxy IP addresses on the cloud which can bring down even a major website.

There are some other parameters (`https://webmasters.googleblog.com/2019/07/a-note-on-unsupported-rules-in-robotstxt.html`) that aren't supported by Google but still respected by other crawlers, but we will not go into them as it's not very important for our purposes.

The code block in Listing 2-16 shows how to parse robots.txt file using robotparser from the urlparse library.

Listing 2-16. Parsing robots.txt

```
# final robot parser code
from urllib import robotparser
from tld import get_tld

def get_rb_object(url):
    robot_url = get_robot_url(url)
    rp = robotparser.RobotFileParser()
    rp.set_url(robot_url)
```

```
    rp.read()
    return(rp)

def parse_robot(url,rb_object):

    flag = rb_object.can_fetch("*", url)
    try:
        crawl_d = rb_object.crawl_delay("*")
    except Exception as E:
        crawl_d = None
    return flag, crawl_d

def get_robot_url(url):
    res = get_tld(url, as_object=True)
    final_url = res.parsed_url.scheme+'://'+res.parsed_url.netloc+'/robots.txt'
    return(final_url)
```

Status codes and retries

When you send an HTTP request, you get back a status code that indicates whether the request has been successful or not. A 200 code indicates success in fetching the page, whereas a 3XX, 4XX, and 5XX refer to redirection, client, or server error. The specific code will give you more information, for example, a 307 indicates a temporary redirect, whereas a 301 means that the content has moved permanently.

A good crawler will check for status codes and will retry requests when codes 4XX or 5XX are raised. The requests library already has "allow redirect" parameter set to true, and so it's supporting that out of the box; you can view the redirected URL as part of the Response.history list.

Crawl depth and crawl order

When you start from an initial seed URL, you can say that you are at depth 0, and all links found on that page which are one click away from it are known as depth 1. Now the links found on pages with depth 1 will be depth 2 since it will take two clicks for us to get to them from the initial seed page. This depth is known as crawl depth, and all production crawlers employ some form of a depth metric so that you don't scrape too many less useful pages.

Now that we know about crawl depth, let's define a new parameter called topN for maximum pages crawled per depth. So now we can get total pages to be crawled by simply multiplying topN with crawl depth.

Crawl order in such an implementation is based on a queue, first in first out (FIFO), and called breadth-first searching. If instead you use a stack, last in first out (LIFO), then you are performing a depth-first search; this is default on a crawling framework called Scrapy which you will see in later chapters. There are some crawlers out there which use a "greedy" approach and fetch whichever pages are fastest to send the response back using various adaptive algorithms. They also continuously monitor response times and make adjustments on crawl delay to account for the slowing of servers.

The other take-home message is that crawling can be easily parallelized onto multiple threads and processes and even distributed among servers by maintaining one single queue which can keep track of pages to be crawled and pages already seen by crawlers and also maintain the persistence of data. This is much easier than you think once you learn about in-memory data structures such as Redis and SQL databases which can provide disk persistence.

Instead of worrying about all that as well as trying to implement a definite crawl order in our function, we will stick with the pop method of the Python set, but I just wanted to put this information out there so that you are familiar with these terms when we do use them in our implementations in the next chapters.

Link importance

An ideal crawling program doesn't try to exhaustively visit all the links it encounters all the time (e.g., a traditional breadth-first searching); rather it incorporates some algorithm to assign a score to each link to determine relative link importance that can guide how frequently a particular page must be recrawled to maintain freshness. There are many algorithms which do that such as Adaptive On-Line Page Importance Computation (OPIC) (www2003.org/cdrom/papers/refereed/p007/p7-abiteboul. html). In some use cases, we are crawling to find similar content to a gold standard document; in such cases, you want to base the link importance score to pages with high similarity (cosine or other metrics) to your gold standard document.

For now, we will simply use a counter method from the collections library to report the number of times a particular link was seen by our crawler; we are still hitting those links only once, but this just provides a rough estimate on the relative importance of each internal link.

Advanced link crawler

Let us incorporate some of the features talked about here to our basic link crawler
(Listing 2-17). We are also elegantly handling the insertion of base_url for relative links.

This is by no means a perfect crawling function, because it still doesn't do much
parsing besides gathering links, and it still doesn't include advanced support for
matching filetypes and so on using regular expressions which you will learn in Chapter 4.

Listing 2-17. Advanced link crawler

```
# final code for advanced crawler

import requests
from bs4 import BeautifulSoup
from tld import get_fld
from tld import get_tld
import time
from collections import Counter

def advanced_link_crawler(seed_url, max_n = 5000):
    my_headers = {
    'User-Agent': 'Mozilla/5.0 (Windows NT 10.0; Win64; x64) AppleWebKit/
    537.36 ' + ' (KHTML, like Gecko) Chrome/61.0.3163.100Safari/537.36'
    }
    initial_url_set = set()
    initial_url_list = []
    seen_url_set = set()
    base_url = 'http://www.'+ get_fld(seed_url)

    res = get_tld(seed_url, as_object=True)
    domain_name = res.fld

    initial_url_set.add(seed_url)
    initial_url_list.append(seed_url)

    robot_object = get_rb_object(seed_url)
    flag, delay_time = parse_robot(seed_url,robot_object)
```

```python
    if delay_time is None:
        delay_time = 0.1

    if flag is False:
        print('crawling not permitted')
        return(initial_url_set, seen_url_set)

    while len(initial_url_set)!=0 and len(seen_url_set) < max_n:
        temp_url = initial_url_set.pop()

        if temp_url in seen_url_set:
            continue
        else:
            seen_url_set.add(temp_url)

            time.sleep(delay_time)

            r = requests.get(url = temp_url, headers = my_headers)
            st_code = r.status_code
            if st_code != 200:
                time.sleep(delay_time)
                r = requests.get(url = temp_url, headers = my_headers)
                if r.status_code != 200:
                    continue
            #print(st_code)
            html_response = r.text
            soup = BeautifulSoup(html_response)
            links = soup.find_all('a', href=True)
            for link in links:
                #print(link['href'])
                if ('http' in link['href']):
                    if domain_name in link['href']:
                        final_url = link['href']
                    else:
                        continue

                elif [char for char in link['href']][0] == '/':
                    final_url = base_url+link['href']
```

```
                # insert url normalization
                #print(final_url)
                final_url = get_normalized_url(final_url)
                flag, delay = parse_robot(seed_url,robot_object)
                # insert robot file checking
                if flag is True:

                    initial_url_set.add(final_url.strip())
                    initial_url_list.append(final_url.strip())
    counted_dict = Counter(initial_url_list)
    return(initial_url_set, counted_dict)
seed_url = 'http://www.jaympatel.com'
advanced_link_crawler(seed_url)
#output
(set(),
 Counter({'http://jaympatel.com': 20,
          'http://jaympatel.com/2018/11/get-started-with-git-and-github-in-
          under-10-minutes': 55,
          'http://jaympatel.com/2019/02/introduction-to-natural-language-
          processing-rule-based-methods-name-entity-recognition-ner-and-
          text-classification': 30,
          'http://jaympatel.com/2019/02/introduction-to-web-scraping-in-
          python-using-beautiful-soup': 29,
          'http://jaympatel.com/2019/02/natural-language-processing-
          nlp-term-frequency-inverse-document-frequency-tf-idf-based-
          vectorization-in-python': 29,
          'http://jaympatel.com/2019/02/natural-language-processing-nlp-
          text-vectorization-and-bag-of-words-approach': 35,
          'http://jaympatel.com/2019/02/natural-language-processing-nlp-
          word-embeddings-words2vec-glove-based-text-vectorization-in-
          python': 32,
          'http://jaympatel.com/2019/02/top-data-science-interview-
          questions-and-answers': 48,
          'http://jaympatel.com/2019/02/using-twitter-rest-apis-in-python-
          to-search-and-download-tweets-in-bulk': 30,
```

```
'http://jaympatel.com/2019/02/why-is-web-scraping-essential-and-
who-uses-web-scraping': 28,
'http://jaympatel.com/2020/01/introduction-to-machine-learning-
metrics': 29,
'http://jaympatel.com/about': 19,
'http://jaympatel.com/books': 39,
'http://jaympatel.com/categories': 93,
'http://jaympatel.com/consulting-services': 21,
'http://jaympatel.com/cv': 2,
'http://jaympatel.com/pages/CV.pdf': 1,
'http://jaympatel.com/tags': 114,
'http://www.jaympatel.com/assets/DoD_SERDP_case_study.pdf': 1}))
```

We will see a lot more crawlers in this book, but for now this is sufficient to understand what a bare minimum production crawler looks like and what it still needs before unleashing it on the wider Web and incurring lots of money in computational time and storage.

Getting things "dynamic" with JavaScript

Let's work through a specific example of where JavaScript makes traditional web scraping impossible.

We are trying to scrape information from a table on the US Food and Drug Administration (US FDA) warning letters database page.

Warning letters are official letters from the US FDA to regulated companies in food, pharmaceutical, and medical device areas which typically discuss regulatory oversights and violations on the part of companies which are discovered by an on-site US FDA inspection of the facilities. Penalties specified in the letter can be as harsh as a complete ban on manufacturing which will obviously have an adverse impact on the company's financial performance, and hence getting this letter is newsworthy for many publicly listed companies.

It looks like a normal HTML table enclosed by table tags with each row enclosed by <tr> and each cell by <td> (see Figure 2-6).

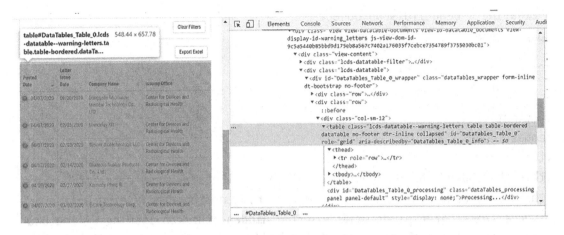

Figure 2-6. *Example of a JavaScript table*

We know enough Beautiful Soup by now to know that all we need is a find_all() call on tr to identify all table rows on a page (Listing 2-18).

Listing 2-18. Scraping the US FDA without executing JavaScript

```
# scraping US FDA without executing JavaScript

import requests
from bs4 import BeautifulSoup
import time

my_headers = {
'User-Agent': 'Mozilla/5.0 (Windows NT 10.0; Win64; x64) AppleWebKit/537.36
' + ' (KHTML, like Gecko) Chrome/61.0.3163.100Safari/537.36'
}

test_url = 'https://web.archive.org/web/20200406193325/https://www.fda.gov/
inspections-compliance-enforcement-and-criminal-investigations/compliance-
actions-and-activities/warning-letters'

r = requests.get(url = test_url, headers = my_headers)
print("request code: ", r.status_code)
html_response = r.text
time.sleep(30)
```

```python
# creating a beautifulsoup object
soup = BeautifulSoup(html_response)

for tr in soup.find_all('tr'):
    print(tr)

# output
request code:   200
<tr>
<th>path hidden</th> <th>Posted Date</th> <th>Letter Issue Date</th>
<th>Company Name</th> <th>Issuing Office hidden</th> <th>Issuing Office
hidden</th> <th>Recipient Office Office hidden</th> <th>Issuing Office old
hidden</th> <th>Letter Type condition hidden</th> <th>Issuing Office</th>
<th>State hidden</th> <th>State hidden</th> <th>Regulated Product hidden</
th> <th>Subject</th> <th>Topics hidden</th> <th>Topics and regulated
Product combined hidden</th> <th>Year hidden</th> <th>Letter Type</
th> <th>Response Letter</th> <th>Closeout Letter</th> <th>checkresponse
hidden</th> </tr>
```

Surprisingly, we are unable to get any information even though visually we can see that the table has lots of nonempty fields. Many newbies at this point think that this may mean that HTML hasn't had time to parse, and putting in a delay between getting the request and parsing it might make a difference. Well, they are on the right track that many web pages load resources asynchronously after the page itself has been loaded, but that isn't what's happening here. Instead, the HTML table is created dynamically using JavaScript and we will only be able to scrape that if we can somehow execute JavaScript like our browsers.

The website in Listing 2-18 is hardly an exception; in fact, all modern websites rely extensively on JavaScript to add dynamic elements to their site, and it will be very hard to understand a web page without knowing more about JavaScript and one of its most popular libraries called jQuery. Please use a console such as http://jsbin.com to work through this section's example code.

There are two ways to insert JavaScript in an HTML page; you can place it between <script> tags in the <head> like shown here:

```html
<head>
    <script>
    some JavaScript code;
    </script>
</head>
```

Alternatively, you can place JavaScript in a separate file and load it through the script tag:

```
<script type="text/JavaScript" src="https://ajax.googleapis.com/ajax/libs/
jquery/1.12.4/jquery.min.js"></script>
```

Many years ago, it was recommended that scripts should always be placed in <head>, but as browsers render pages from top to bottom, that resulted in slower page loads which adversely impacts search engine rankings and user engagement. Another issue was that if you try to apply style to HTML elements such as body before that has loaded, then nothing would happen if you don't explicitly mention in the code to wait for the body to load.

So it has become a norm to place scripts at the bottom of the page just before closing the <body> tag. Some scripts (e.g., Google Analytics) should almost always be placed in head because you want to record a user session even if the user leaves your site before the entire page is loaded.

Mentioning script attributes is not mandatory, and this is a totally valid script too:

```
<script src="https://ajax.googleapis.com/ajax/libs/jquery/1.12.4/jquery.
min.js"></script>
```

Single-line comments can be made by // whereas multiline comments start and end with " */ " and " */ ", respectively.

Variables and data types

Variables are initialized by appending them with "var"; you do not need to initialize a variable before declaring it, and in that case, the default value will be set to undefined.

JavaScript has all the variable data types as Python, and it has almost the same syntax too; lists and dictionaries are called arrays and objects, respectively, in JavaScript parlance.

A new empty array can be initialized as

```
var new_array = new Array();
```

If you want to initialize a filled array, that's pretty intuitive:

```
var new_array = [element1, element2..];
```

The length of an array can be found by .length property:

```
var array_length = new_array.length
```

Similarly, JavaScript objects can be initialized and filled similar to a Python dictionary:

```
var person = new Object();
//var person = {} is valid too
person.firstName = "Jay";
person.lastname = "Patel";
// person["lastname"] is valid too
```

Functions

JavaScript functions can be initialized by appending a function variable and enclosing the code in curly brackets (Listing 2-19).

Listing 2-19. JavaScript functions

```
function initialize_person(parameter1, parameter2) {
var person = {};
person.firstName = parameter1;
person.lastname = parameter2;
return person;
}
initialize_person("Jay", "Patel")
// Output:
[object Object] {
  firstName: "jay",
  lastname: "Patel"
```

You can initialize an anonymous function in JavaScript; the only condition is that it must be assigned to a variable. Alert() simply prints that in your screen:

```
var msg = function(firstName) {
    alert("Hello " + firstName);
};
```

Conditionals and loops

JavaScript has if and else if just like Python, with only slight changes to syntax which includes wrapping the condition in parentheses and the statements in curly brackets. All the operators such as "<" can be used; the only subtle difference is that the JavaScript operator for checking if two values are the same or not is "===". Note: "==" is a valid operator in JavaScript, but it does type conversions before evaluating the condition, so if you are comparing a string and an integer, using == may result in true if the underlying value is the same even though that is not usually expected (Listing 2-20).

Listing 2-20. JavaScript conditional statements

```
if(condition)
{
    // code to be executed if condition is true
}
else{
    //Execute this code..
}
using else if for multiple conditions
if(condition)
{
    // code to be executed if condition is true
}
else if(condition)
{
    // code to be executed if condition is true
}
else if(condition)
{
    // code to be executed if condition is true
}
```

JavaScript also provides for another conditional statement called switch which is pretty similar to if…elif in usage.

There are two loops available in JavaScript, the for loop and while loop; pseudocode is shown in the following. This will look very familiar to you in case you already know C and related languages (see Listing 2-21).

71

Listing 2-21. For and while loops

```
for(counter_initializer; condition; iteration_counter)
{
    // Code to be executed
}
// example
for(i = 0; i<10; i++)
{
    // Code to be executed
}
while(condition)
{
    // code executed till  condition is true
}
```

HTML DOM manipulation

The Document Object Model (DOM) treats HTML as a tree structure, and interacting with it via traversal, insertion, and manipulation of new elements, styles, and so on is a key aspect of learning how JavaScript interacts with web pages. The predominant library used for it is called jQuery so you need to be somewhat familiar with its syntax as well as plain JavaScript APIs.

Let us take a very simple use case of selecting an element by id and changing its color. As you can see, the jQuery syntax is much easier to understand, and this makes it far more common (Listing 2-22).

Listing 2-22. JavaScript DOM manipulation example

```
// plain JavaScript
    function changeColor() {
    var msg;
    msg = document.getElementById("first");
    msg.style.color = "green";
    }
```

```
// JQuery version
function changeColor() {
    var msg;
// msg = $(document).$('#first');
    msg = $('#first');
    msg.css("color", "green");
    }
```

Let's edit the same HTML code we used earlier in this chapter, except to add a clickable button and the script to dynamically insert styling (Listing 2-23). Save the code as HTML and open it in Chrome.

Listing 2-23. Sample HTML page with JavaScript

```
<!DOCTYPE html>
<html>
<script>
    function changeColor() {
    var msg;
    msg = document.getElementById("firstHeading");
    msg.style.color = "green";
    }
</script>
<body>

<h1 id="firstHeading" class="firstHeading" lang="en">Getting Structured
Data from the Internet:</h1>

<h2>Running Web Crawlers/Scrapers on a Big Data Production Scale</h2>

<p id = "first">
Jay M. Patel
</p>

<input type="button" value="Change color!" onclick="changeColor();">

</body>

</html>
```

It will look like Figure 2-7 before clicking the button.

Figure 2-7. *Example page before triggering JavaScript*

On clicking the button, the JavaScript function changeColor is triggered, and we will have the page as shown in Figure 2-8.

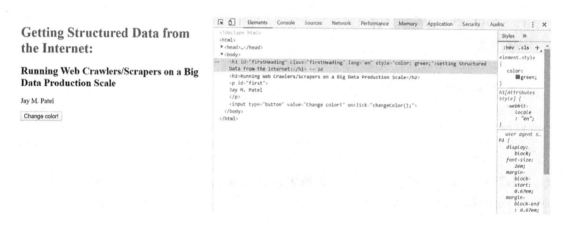

Figure 2-8. *Page after triggering JavaScript*

Notice that the font color of <h1> is now changed, and style= "color:green" is dynamically inserted into the HTML DOM.

AJAX

AJAX stands for Asynchronous JavaScript and XML and refers to a set of methods for fetching the data using GET/POST HTTP calls on the client side without reloading the entire web page.

Even though XML is mentioned in AJAX, in reality we use JSON files to request data from most websites. POST calls with xhr methods assume that content-type is text/plain, whereas for jQuery we are required to explicitly mention it (Listing 2-24).

Listing 2-24. AJAX example

```
// GET calls
var xhr = new XMLHttpRequest();
xhr.open('GET', '/api/name');
xhr.onload = function some_function(parameter) {
//do something
}

//Jquery version

$.get('/api/name').then(
function some_function(parameter) {
//do something
}

// POST calls
var xhr = new XMLHttpRequest();
xhr.open('POST', '/api/name');
xhr.send('data1');

// JQuery version
$.ajax({
method: 'POST',
url: '/api/name',
contentType: 'text/plain',
data: 'data1'
 });
```

Going through AJAX works in detail will at least take a dozen pages and is out of the scope for this book, but I wanted to put it here so that you can conceptually understand how that US FDA page we talked about earlier was able to fetch data to populate the table after the web page itself was loaded.

Scraping JavaScript with Selenium

Selenium is a powerful Java-based library originally intended for automated website testing with most browsers such as Chrome and Firefox. It provides bindings to popular languages such as Python, Ruby, and so on, so you can use it as a library as part of your workflow for executing JavaScript before scraping the information using BeautifulSoup.

You can download Selenium from its website or using pip. In addition to that, you will also need a browser-specific webdriver which can work with Selenium. If you use Chrome, then just go to the chromedriver (`https://chromedriver.chromium.org/`) site and download the version most appropriate for your Google Chrome version; the latest version at the time of writing this book is 81.0.4044.69.

The webdriver object is pretty similar to Beautiful Soup in syntax and provides a lot of flexibility in selecting elements from the page using tag names, XPath, id, class names, and so on. You can also interact with any elements of the page such as forms, buttons, click boxes, and so on using the select method of the webdriver API.

You can get only the first instance of the webelement object by using the get_element method, and to get all of the matching instances, you can use the get_elements method (Listing 2-25).

Listing 2-25. Selenium example

```
from selenium import webdriver
from selenium.webdriver.common.by import By
import time

test_url = 'https://web.archive.org/web/20200406193325/https://www.fda.gov/
inspections-compliance-enforcement-and-criminal-investigations/compliance-
actions-and-activities/warning-letters'

option = webdriver.ChromeOptions()
option.add_argument("--incognito")
chromedriver = your_chromedriver_location_on local_filesystem
browser = webdriver.Chrome(chromedriver, options=option)
browser.get(test_url)

time.sleep(15)
```

```
print(browser.find_element_by_tag_name('h1').text)
element_list = browser.find_elements_by_tag_name('h1')
for element in element_list:
    print(element.text)

browser.close()
# output

U.S. Food and Drug Administration
U.S. Food and Drug Administration
Warning Letters
```

Scraping the US FDA warning letters database

We are going to scrape from a snapshot of the US FDA warning letters database using Selenium to execute JavaScript (see Figure 2-9).

04/08/2020	04/07/2020	CBD Online Store	Center for Drug Evaluation and Research | CDER	Unapproved Products Related to the Coronavirus Disease 2019 (COVID-19)		
04/08/2020	04/08/2020	Genesis 2 Church	Center for Drug Evaluation and Research | CDER	Unapproved Products Related to the Coronavirus Disease 2019 (COVID-19)		
04/07/2020	09/20/2019	Dongguan Microview Medical Technology Co., LTD	Center for Devices and Radiological Health	CGMP/QSR/Medical Devices/Adulterated		
04/07/2020	02/03/2020	Mandelay Kft	Center for Devices and Radiological Health	CGMP/QSR/Medical Devices/Adulterated		
04/07/2020	02/03/2020	Steiner Biotechnology, LLC	Center for Devices and Radiological Health	Good Laboratory Practice (GLP) for Nonclinical Laboratory Studies		
04/07/2020	02/14/2020	Okamoto Rubber Products Co., Ltd.	Center for Devices and Radiological Health	CGMP/QSR/Medical Devices/Adulterated		
04/07/2020	02/27/2020	Kennedy, Philip R.	Center for Devices and Radiological Health	Investigational Device Exemptions (Clinical Investigator)		
04/07/2020	03/02/2020	E-Care Technology Corp.	Center for Devices and Radiological Health	CGMP/QSR/Medical Devices/Adulterated		
04/07/2020	03/18/2020	Cafe Valley Bakery, LLC	Office of Human & Animal Food Operations East 6	CGMP/Food/Prepared, Packed or Held Under Insanitary Conditions/Adulterated		

Showing 1 to 10 of 2,863 entries

Previous 1 2 3 4 5 ... 287 Next

Figure 2-9. *US FDA table*

Our goal is to scrape information from all the 287 pages of the US FDA warning letters database. Before we dive in and start writing any code, let us go through it conceptually and imagine all the steps we will need to do it.

- We need Selenium to load the page and use Beautiful Soup to scrape ten rows from the table.

- Later, we need Selenium to click the next button and so on and keep using BeautifulSoup to scraping rows off the page once it's loaded.

- This is easy enough since we can manually figure out the XPath of both the first page (//*[@id="DataTables_Table_0_paginate"]/ul/li[2]/a) and the next button (//*[@id="DataTables_Table_0_next"]/a).

But we should ask if this is indeed the most computationally efficient way to do things. For some pages, pagination through JavaScript may be the only option, but let us explore other elements on the page before we embark on this time-consuming method (Figure 2-10).

Posted Date ▾	Letter Issue Date ⇕	Company Name ⇕	Issuing Office ⇕	Subject ⇕	Response Letter	Show 10 ▾ entries
						10
04/08/2020	04/07/2020	Savvy Holistic Health dba Holistic Healthy Pet	Center for Drug Evaluation and Research | CDER	Unapproved Products Related to the Coronavirus Disease 2019 (COVID-19)		25 ut
						50
04/08/2020	04/07/2020	CBD Online Store	Center for Drug Evaluation and Research | CDER	Unapproved Products Related to the Coronavirus Disease 2019 (COVID-19)		100
04/08/2020	04/08/2020	Genesis 2 Church	Center for Drug Evaluation and Research | CDER	Unapproved Products Related to the Coronavirus Disease 2019 (COVID-19)		All

Export Excel

Figure 2-10. *US FDA table pagination*

If we scroll up to the top of the table in Figure 2-10, we see the Export Excel button. Bingo! This may be everything we need. Unfortunately, after I downloaded it, I discovered that the Excel file doesn't provide URLs to the individual warning letters themselves, and without that we would still need to scrape the URLs from tables.

Next to the Export Excel button, we have the Show entries drop-down list, and here we can simply select "All" to get all the 2800+ rows in one go.

The complete script for this is shown in Listing 2-26.

Listing 2-26. Scraping the US FDA table using Selenium

```
from selenium import webdriver
from selenium.webdriver.common.by import By
from selenium.webdriver.support.ui import Select

import time

from bs4 import BeautifulSoup
import numpy as np
import pandas as pd

test_url = 'https://web.archive.org/web/20200406193325/https://www.fda.gov/
inspections-compliance-enforcement-and-criminal-investigations/compliance-
actions-and-activities/warning-letters'

option = webdriver.ChromeOptions()
option.add_argument("--incognito")

chromedriver = your_chromedriver_location_on local_filesystem
browser = webdriver.Chrome(chromedriver, options=option)
browser.get(test_url)

time.sleep(30)

element = browser.find_element_by_xpath('//*[@id="DataTables_Table_0_
length"]/label/select')

select = Select(element)

# select by visible text
select.select_by_visible_text('All')

posted_date_list = []
letter_issue_list = []
warning_letter_url = []
company_name_list = []
issuing_office_list = []

soup_level1=BeautifulSoup(browser.page_source, "lxml")
```

```
for tr in soup_level1.find_all('tr')[1:]:
    tds = tr.find_all('td')
    posted_date_list.append(tds[0].text)
    letter_issue_list.append(tds[1].text)
    warning_letter_url.append(tds[2].find('a')['href'])
    company_name_list.append(tds[2].text)
    issuing_office_list.append(tds[3].text)
    #print(tds[0].text, tds[1].text,tds[2].find('a')['href'], tds[2].text,
    tds[3].text)
browser.close()
df = pd.DataFrame({'posted_date': posted_date_list,
    'letter_issue': letter_issue_list,
    'warning_letter_url': warning_letter_url,
    'company_name': company_name_list,
    'issuing ofice': issuing_office_list})
df.head()
Output:
```

The output is shown in Figure 2-11.

out[29]:

	company_name	issuing ofice	letter_issue	posted_date	warning_letter_url
0	Savvy Holistic Health dba Holistic Healthy Pet	Center for Drug Evaluation and Research | CDER	04/07/2020	04/08/2020	/inspections-compliance-enforcement-and-crimin...
1	CBD Online Store	Center for Drug Evaluation and Research | CDER	04/07/2020	04/08/2020	/inspections-compliance-enforcement-and-crimin...
2	Genesis 2 Church	Center for Drug Evaluation and Research | CDER	04/08/2020	04/08/2020	/inspections-compliance-enforcement-and-crimin...
3	Dongguan Microview Medical Technology Co., LTD	Center for Devices and Radiological Health	09/20/2019	04/07/2020	/inspections-compliance-enforcement-and-crimin...
4	Mandelay Kft	Center for Devices and Radiological Health	02/03/2020	04/07/2020	/inspections-compliance-enforcement-and-crimin...

Figure 2-11. *Scraped data from the US FDA table*

Scraping from XHR directly

There are two ways we can make the preceding approach even more efficient which can allow us to scrape hundreds of sites from a server. An easy way to save computational resources is by switching to a headless browser like PhantomJS or headless Chrome. Headless browsers will still execute JavaScript but not have a UI, so you save memory which allows you to load up more parallel processes.

But an even better idea is bypassing JavaScript and Selenium altogether and instead hitting on the primary source from where the JavaScript is loading its data.

We have already seen that JavaScript uses GET or POST calls using XMLHttpRequest (XHR) or a jQuery wrapper to get data in the form of JSON, XML, and so on from an undocumented API, and it simply parses and loads it into a table, map, graph, and so on.

Google Chrome has developer tools that easily allow us to inspect traffic and look for JSON files. Once we identify it, we can start hitting the endpoint directly, and that can let us scale up really quickly (see Figure 2-12).

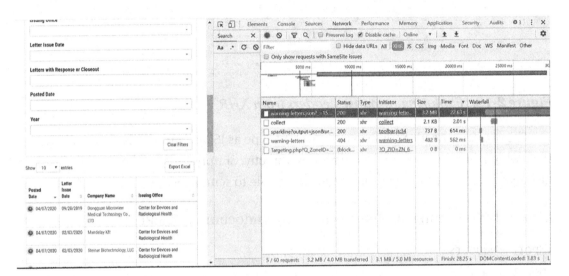

Figure 2-12. *XHR*

So the network tab will show all the XHRs sent after the page was loaded; we already know that our data table takes some time to fully load so the first thing we should do is sort by time. The first request looks pretty promising, so let's click it to get request and response headers (see Figure 2-13).

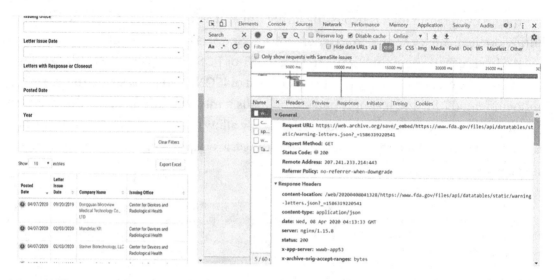

Figure 2-13. *Exploring response headers from XHR*

The response header confirms the content-type as JSON, and we also get a request URL, which we can use to make the request directly. Simply copy this URL shown in Listing 2-27, and let's use Python's request module to send a request.

Listing 2-27. Getting results directly using undocumented API

```
import requests
import numpy as np
import pandas as pd
import io
from bs4 import BeautifulSoup

my_headers = {
'User-Agent': 'Mozilla/5.0 (Windows NT 10.0; Win64; x64) AppleWebKit/537.36
' + ' (KHTML, like Gecko) Chrome/61.0.3163.100Safari/537.36'
}

test_url = 'https://web.archive.org/save/_embed/https://www.fda.gov/files/
api/datatables/static/warning-letters.json?_=1586319220541'

r = requests.get(url = test_url, headers = my_headers)
print("request code: ", r.status_code)
html_response = r.text
```

```python
string_json2 = io.StringIO(html_response)
df = pd.read_json(string_json2)
def get_abs_url(html_tag):
    soup = BeautifulSoup(html_tag,'lxml')
    abs_url = 'https://www.fda.gov' + soup.find('a')['href']
    company_name = soup.find('a').get_text()
    return abs_url, company_name
df["abs_url"], df["company_name"] = zip(*df["field_company_name_warning_
lette"].apply(get_abs_url))
df.to_csv("warning_letters_table.csv")
df.head()
#output
```

The output is shown in Figure 2-14.

field_building	field_change_date_2	field_company_name_warning_lette	field_detailed_description_2	field_issuing_office	field_issuing_office_loc_name	field_letter_
Philadelphia District Office	01/13/2015	<a href="/inspections-compliance-enforcement-a...	Low Acid Dog and Cat Canned Food Regulation/Ad...		Philadelphia District Office	
Center for Devices and Radiological Health	01/19/2015	<a href="/inspections-compliance-enforcement-a...	Medical Device Reporting/Misbranded		Center for Devices and Radiological Health	
Center for Tobacco Products	01/19/2015	<a href="/inspections-compliance-enforcement-a...	Family Smoking Prevention and Tobacco Control ...		Center for Tobacco Products	
New York District Office	01/19/2015	<a href="/inspections-compliance-enforcement-a...	CGMP/Dietary Supplement/Adulterated /Misbranded		New York District Office	
Dallas District Office	01/19/2015	<a href="/inspections-compliance-enforcement-a...	Illegal Drug Residue		Dallas District Office	

Figure 2-14. *Data from the undocumented API*

It seems like we have got data including a lot more hidden fields for all 2800+ entries with just one request call without messing around with pagination, Selenium, and so on. We used a helper function to read the <a> tags from "field_company_name_warning_lette" column and split the information into the absolute URLs and company names.

I hope you've gotten a fair idea of how to scrape individual web pages, crawl an entire site, and lastly find your way around dynamic content created by JavaScript using Selenium as well as directly calling the XHR URLs.

Summary

We learned the basics of modern web pages comprising of HTML, CSS, and JavaScript and scraped structured information from them using popular Python libraries such as Beautiful Soup, lxml, and Selenium. We demonstrated its practical usage by scraping the US FDA warning letters database.

We will introduce cloud computing in Chapter 3 and see how we can leverage Amazon Web Services (AWS) to perform web crawling at scale.

CHAPTER 3

Introduction to Cloud Computing and Amazon Web Services (AWS)

In this chapter, you will learn the fundamentals of cloud computing and get an overview of select products from Amazon Web Services. AWS offers a free tier where a new user can access many of the services free for a year, and this will make almost all examples here close to free for you to try out. Our goal is that by the end of this chapter, you will be comfortable enough with AWS to perform almost all the analysis in the rest of the book on the AWS cloud itself instead of locally.

However, if you plan to work through all the examples in this book locally on your personal computer or at an on-premises (on-prem) server, then this chapter might feel redundant. In that case, you can pick and choose the sections described as follows as per your requirements:

- IAM and EC2 sections if you want to replicate the Listing 4-6 example from Chapter 4.

- The IAM section for setting up PostgreSQL on Amazon RDS for Chapter 5 examples.

85

© Jay M. Patel 2020
J. M. Patel, *Getting Structured Data from the Internet*, https://doi.org/10.1007/978-1-4842-6576-5_3

- IAM and S3 sections are necessary for Chapters 6 and 7 since we will be using data compiled by a nonprofit called common crawl which is only publicly available on S3 through AWS open registry. You will have to be somewhat comfortable with the Python SDK library (Boto3) and S3 described in this chapter even if you plan to process that data locally. I recommend that you should also go through the EC2 section since ideally computations should be performed as near to where the data is stored to save on time taken for downloading data. For example, a typical Internet speed of 30 MBPS will take about 5 minutes to download 1 GB worth of data from S3. This means that if you want to run some fast algorithms on your dataset such as regular expressions discussed in Chapter 4, then your limiting factor is in fact the bandwidth, and it would've been more efficient if you performed the computations on an AWS-based server such as EC2 itself which could give you a bandwidth of 5–10 GB/s if the EC2 server and S3 data were both located in the same geographical region.

- If you plan on working through examples in Chapters 7 and 8 for performing distributed computing using multiple servers with all the bells and whistles of starting/stopping servers automatically, then you need to read the entirety of this chapter including SNS and SQS sections.

There are lots of good reasons why you should be using cloud computing for web scraping; the top reason is that it prevents the risk of your local computer's IP address being blacklisted by a popular website because your crawler tried to fetch too many pages from it.

For example, about a year ago, a mid-sized client of mine got a rude shock when none of their ~500 employees could get any search results from Google. It turns out that one of the new employees there somehow managed to get their IP addresses blacklisted by trying to aggressively scrape Google's results pages. In cases like this, it's safer to run crawlers through an AWS server or at the very least use an IP rotation service (discussed in Chapter 8) if you have to do it locally.

What is cloud computing?

Cloud computing is the on-demand availability of computer system resources such as data storage and computing power without direct active management by the user.

Cloud computing is the one powering all the well-known Internet apps of today including Airbnb and Netflix. Almost all of the applications and use cases mentioned in Chapter 1 can be implemented by using a combination of cloud computing services. A full list of AWS current customers and the specific services they are using is available on the AWS case studies page (`https://aws.amazon.com/solutions/case-studies`).

We practice what we preach, and all the current products at Specrom Analytics such as our text analytics APIs or historical news APIs are completely being served by cloud servers.

List of AWS products

AWS has products which cover almost all aspects of serving content over the Internet to your customer, so it is very likely that you can run your entire technology stack on AWS products; however, one thing that stumbles many users is predicting the total cost of AWS usage for a particular application.

All AWS products have associated storage, instance usage and provisioning, and movement of data cost. The official AWS Cost Explorer helps some in this regard, but it's still not enough, and this is why many people have built a successful consultancy career out of it. We will try and provide a rough estimate of pricing wherever possible so that you do not get any unexpected shock by running any examples seen in the book.

There are entire books dedicated to learning about all AWS products so we cannot go through all of those in this chapter, but here we will focus our attention on most important products (IAM, AWS Management Console, EC2, S3, SNS, SQS) which should give you an idea on compute, data storage, management, and security, identity, and compliance capabilities of AWS.

In the next chapters, we will introduce SQL and NoSQL databases and present a selection of AWS products of particular interest to us.

a. **Security, identity, and compliance**: AWS Identity and Access Management (IAM)

b. **Storage**: Amazon S3, Amazon Elastic Block Store (EBS)

  c. **Compute**: Amazon Elastic Compute Cloud (EC2), AWS Lambda

  d. **Search**: Amazon CloudSearch, Amazon Elasticsearch Service

  e. **Management tools**: AWS Management Console, Amazon CloudWatch, AWS CloudFormation, AWS CloudTrail

  f. **Database**: Amazon DocumentDB, Amazon Athena, Amazon Aurora, Amazon Relational Database Service (RDS)

  g. **Analytics**: AWS Glue, Amazon Elastic MapReduce

  h. **Machine learning**: Amazon Textract, Amazon Translate

  i. **Messaging**: Amazon Simple Email Service (Amazon SES), Amazon Simple Notification Service (SNS), Amazon Simple Queue Service (SQS)

  j. **Application services**: Amazon API Gateway

  k. **Networking and content delivery**: Amazon CloudFront

How to interact with AWS

There are four main ways to interact with AWS products:

1. **AWS Management Console**: It's a web-based management application with a user-friendly interface and allows you to control and manipulate a wide variety of AWS resources. We will use this as our main way to interact with AWS.

2. **AWS Command Line Interface (CLI)**: This is a downloadable tool which allows you to control all aspects of AWS with your computer's command-line interface. Since this operates from the command line, you can easily automate any or all aspects through scripts.

3. **AWS SDKs**: AWS already has software development kits (SDKs) for programming languages such as Java, Python, Ruby, C++, and so on. You can easily use these to tightly integrate AWS services such as machine learning, databases, or storage (S3) with your application.

4. **AWS CloudFormation**: AWS CloudFormation allows you to use programming languages or a simple text file to model and provision, in an automated and secure manner, all the resources needed for your applications across all regions and accounts. This is frequently referred to as "Infrastructure as Code," which can be broadly defined as using programming languages to control the infrastructure.

All these four methods have pros and cons with respect to usability, maintainability, robustness, and so on, and there is a lively debate on Stack Overflow (`https://stackoverflow.com/questions/52631623/aws-cli-vs-console-and-cloudformation-stacks`) and Reddit (`www.reddit.com/r/aws/comments/5v2s8d/cloudformation_vs_aws_cli_vs_sdks/`) on which approach is the best. I think for our use case of building a production-ready web crawling system, we will be better off to start off with AWS Management Console and directly jump to the Python AWS SDK library for more complicated steps.

AWS Identity and Access Management (IAM)

IAM allows you to securely control AWS resources. When you register for AWS, the user id you create is called a **root user** which has the most extensive permissions available within IAM.

AWS recommends that you should refrain from using root user id for everyday tasks and instead use it only to create your first IAM user.

Let us break down IAM into its components:

- **IAM user**: Is used to grant AWS access to other people.

- **IAM group**: Grant the same level of access to multiple people.

- **IAM role**: Grant access to other AWS resources. For example, allow the EC2 server to access the S3 bucket.

- **IAM policy**: Used to define the granular level permissions for the IAM user, IAM group, or IAM role.

Setting up an IAM user

Go to the IAM dashboard and pick users from the left pane and click the add user button.

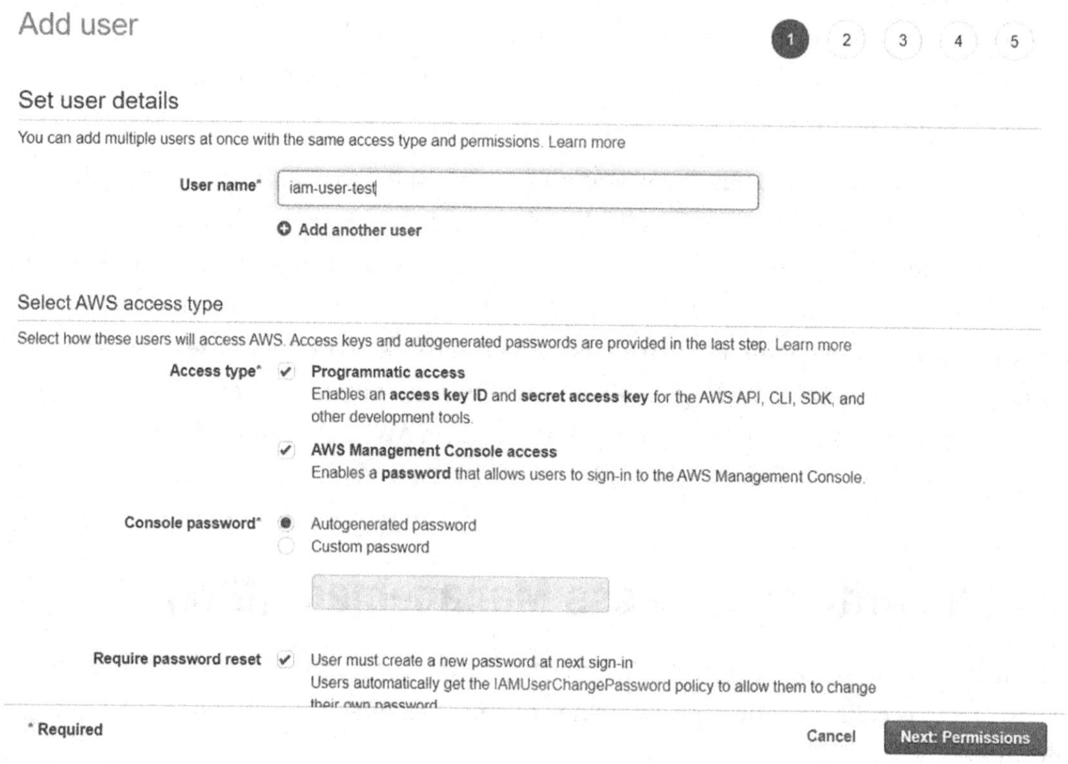

Figure 3-1. *IAM user access types*

For the access type, you should pick both programmatic access and AWS Management Console access so that you can access AWS resources through SDKs as well as through the UI (Figure 3-1). Click next permissions to get the next screen (Figure 3-1).

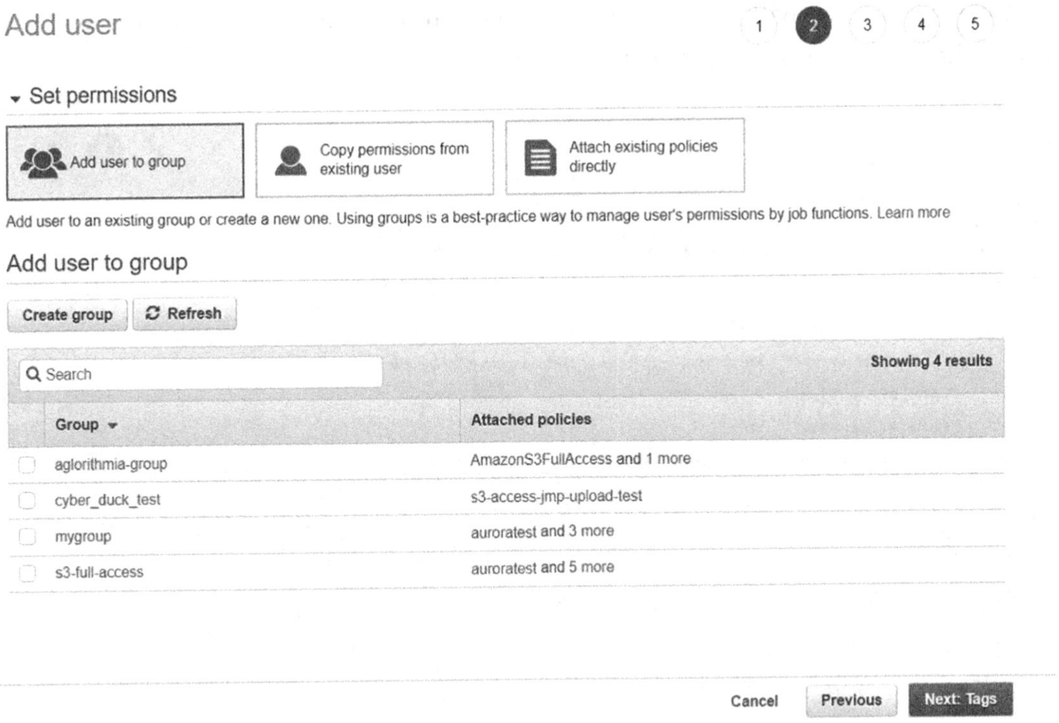

Add user   1 ② 3 4 5

▾ Set permissions

| 👥 Add user to group | 👤 Copy permissions from existing user | 📄 Attach existing policies directly |

Add user to an existing group or create a new one. Using groups is a best-practice way to manage user's permissions by job functions. Learn more

Add user to group

Create group   ↻ Refresh

Q Search   Showing 4 results

Group ▾	Attached policies
☐ aglorithmia-group	AmazonS3FullAccess and 1 more
☐ cyber_duck_test	s3-access-jmp-upload-test
☐ mygroup	auroratest and 3 more
☐ s3-full-access	auroratest and 5 more

Cancel   Previous   Next: Tags

Figure 3-2. *IAM user groups*

If you have not used AWS before, then the screen shown in Figure 3-2 should contain no groups. In that case, just click create group (Figure 3-3). Now, use the search box to search for "AmazonS3FullAccess"; once you find it, click the check box. Similarly, search and tick for "AmazonEC2FullAccess," "AmazonSQSFullAccess," and "AmazonSNSFullAccess."

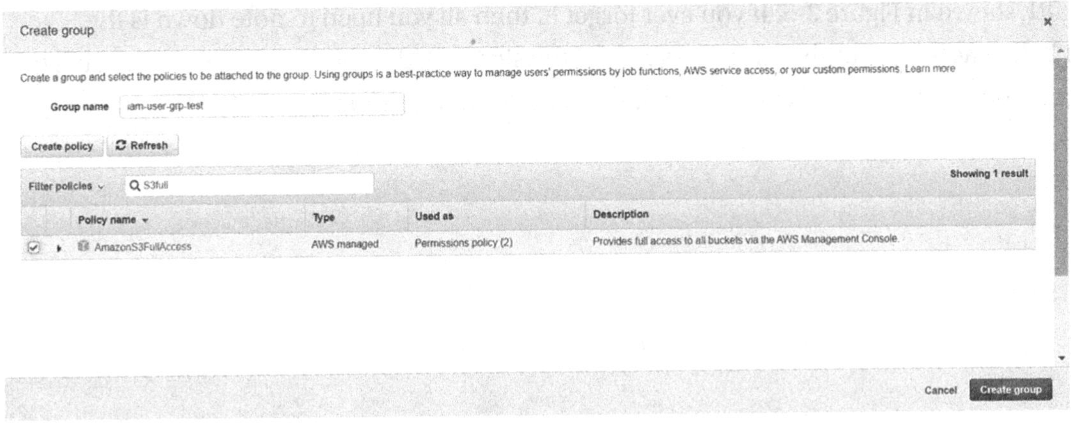

Create group   ✕

Create a group and select the policies to be attached to the group. Using groups is a best-practice way to manage users' permissions by job functions, AWS service access, or your custom permissions. Learn more

Group name   iam-user-grp-test

Create policy   ↻ Refresh

Filter policies ▾   Q S3full   Showing 1 result

Policy name ▾	Type	Used as	Description
☑ ▸ 📋 AmazonS3FullAccess	AWS managed	Permissions policy (2)	Provides full access to all buckets via the AWS Management Console.

Cancel   Create group

Figure 3-3. *IAM group*

Click next on tags without modifying anything. You should see the final review screen shown in Figure 3-4. Just click create user.

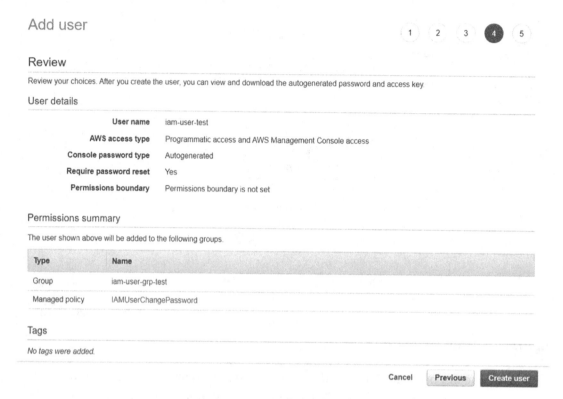

Figure 3-4. *IAM user final review*

You should save the secret access key which can be used for programmatic access and password used for access via AWS Management Console. Note down the sign-in URL shown in Figure 3-5. If you ever forget it, then all you need to note down is the 12-digit account number shown as part of the sign-in URL. You can obviously be able to reset the secret access keys and/or password later if required.

Figure 3-5. *Adding an IAM user*

Using the account number, you can just go to the AWS sign-in screen (`https://signin.aws.amazon.com`) and select the IAM user, enter the account ID, and click next to proceed with entering your IAM username and password. See Figure 3-6.

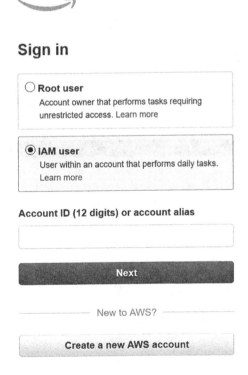

Figure 3-6. *AWS login screen*

Setting up custom IAM policy

You do not have to select IAM policies from a prepopulated list, and AWS gives you the ability to create fine-grained IAM policies with the help of a JSON file.

Let us go through an example where we are assigning permissions to perform read, write, and delete operations on objects located in a particular S3 bucket; we will discuss S3 in the next section, but for now just think of S3 objects as files.

Let's go to the IAM console and click policies; you should get a page shown in Figure 3-7. Click the "create policy" button; you should see a text box with an editable JSON field.

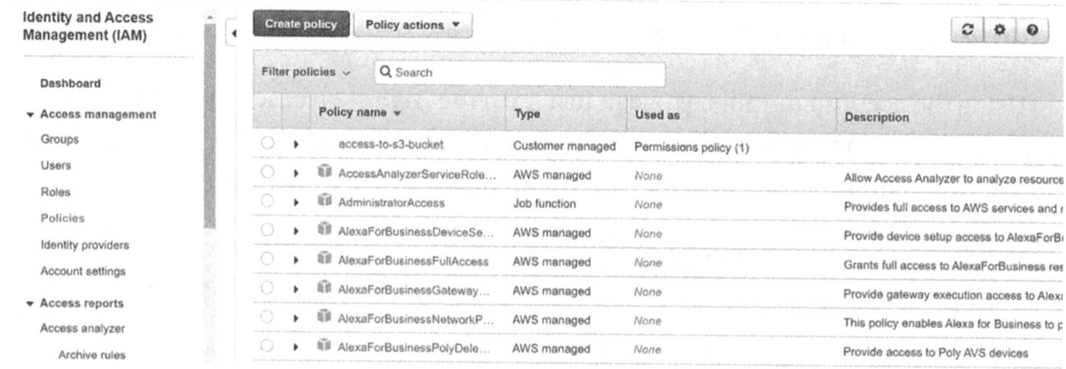

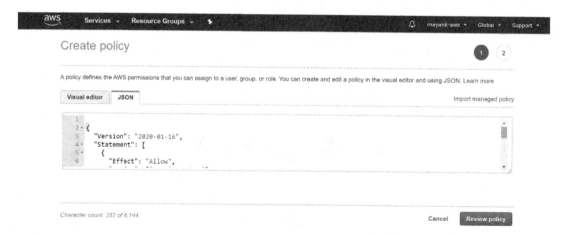

Figure 3-7. *IAM policy list*

You should paste the JSON shown in Listing 3-1 in the text box shown in Figure 3-8.

Figure 3-8. *Create policy*

Listing 3-1. IAM policy

```
{
  "Version": "2012-10-17",
  "Statement": [
    {
      "Effect": "Allow",
      "Action": ["s3:ListBucket"],
      "Resource": ["arn:aws:s3:::ec2-testing-for-s3-permissions"]
    },
    {
      "Effect": "Allow",
      "Action": [
        "s3:PutObject",
        "s3:GetObject",
        "s3:DeleteObject"
      ],
      "Resource": ["arn:aws:s3:::ec2-testing-for-s3-permissions/*"]
    }
  ]
}
```

Once you click the review policy, you'll get the page shown in Figure 3-9 where there is a summary of the new policy, and it prompts you to enter the name and description of the new policy. For our reference, just type "access-to-s3-bucket" as the policy name.

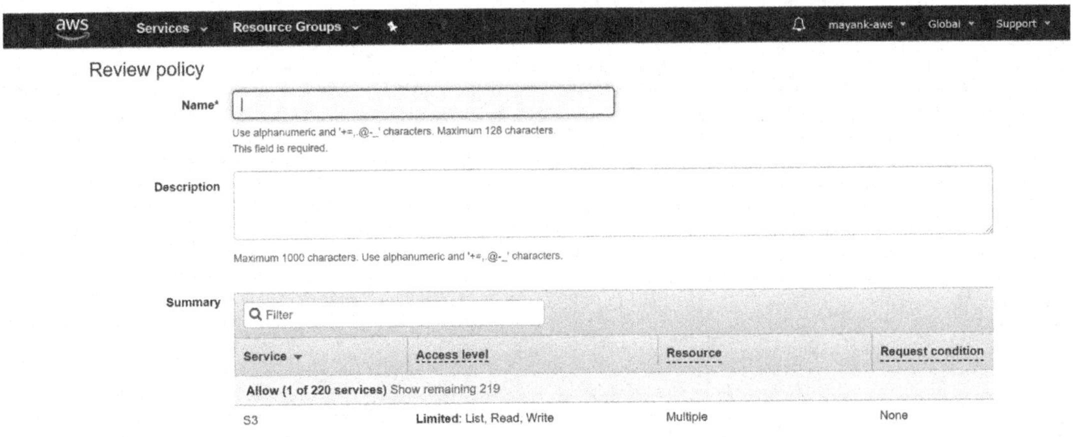

Figure 3-9. *Review policy*

Setting up a new IAM role

In the previous section, you have seen how to create an IAM user, group, and policy, so as a final piece, let's also show you how to set up a new IAM role.

Go to the IAM console and click roles on the left pane to see the screen shown in Figure 3-10. Click "create role" to start the process of setting up a new IAM role.

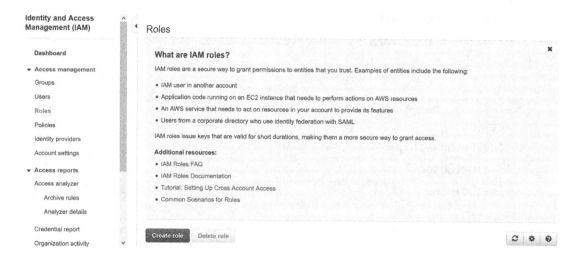

Figure 3-10. *IAM role*

You will be prompted to define what kind of new IAM role you want; for our purposes, let's keep things simple and select the first option (AWS service) and then pick EC2. See Figure 3-11. We will discuss EC2 in the next sections, but for now just think of them as a virtual server.

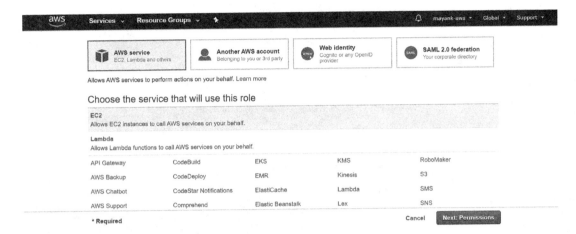

Figure 3-11. *IAM role (cont.)*

Select the policy you just created in the previous section. If you can't find it immediately, then just click filter policies and select customer managed. Click next. See Figure 3-12.

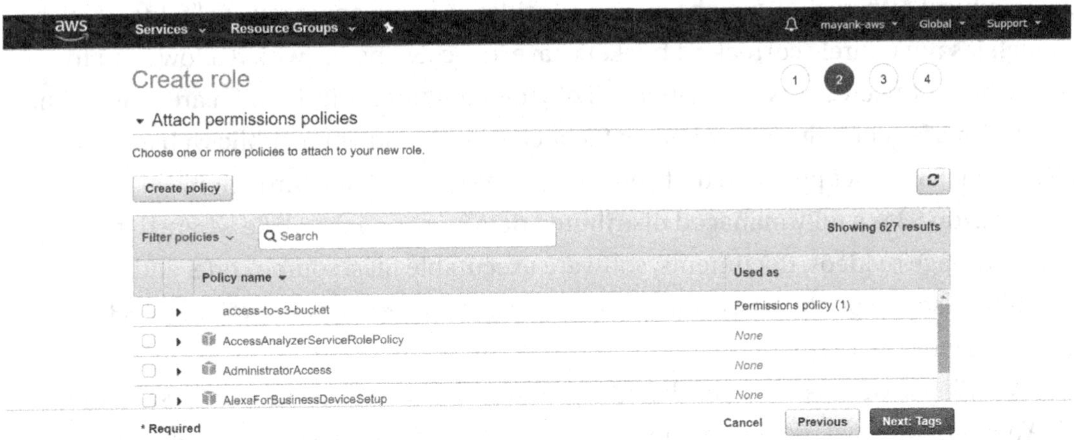

Figure 3-12. *IAM role (cont.)*

Finally, on the last page, just name the role as "ec2_to_s3". See Figure 3-13.

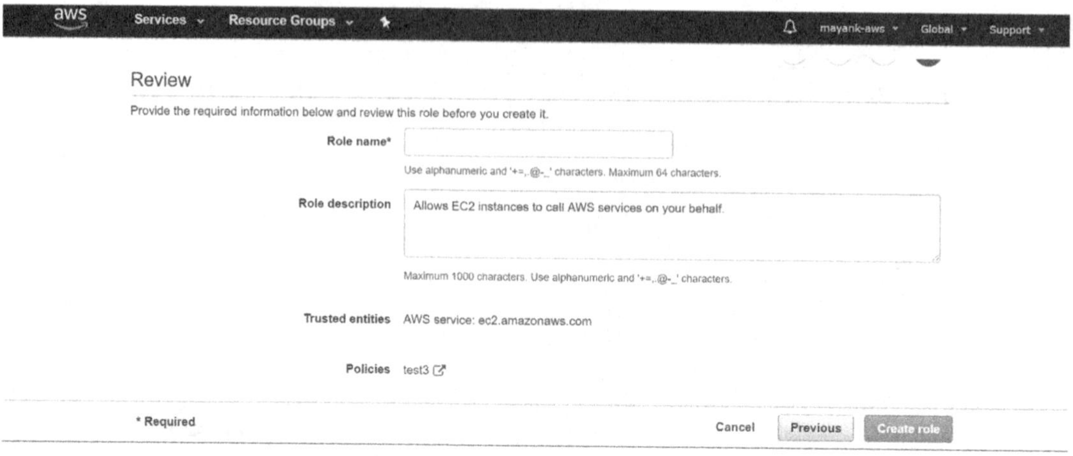

Figure 3-13. IAM role (cont.)

Amazon Simple Storage Service (S3)

S3 objects are stored in constructs called buckets which are created in a specific geographical AWS region, which you can consider as being analogous to "folders" even though it's not entirely correct. S3 buckets have an access policy which allows you to make them publicly accessible, where all objects contained within them are open to the public. This is generally done when an S3 bucket holds data for a public website. You can also set an S3 bucket policy so that you can host a static website through it.

Amazon S3 is a fully managed distributed data store or a data lake where data such as images, videos, documents, software executable files, source code, and almost anything else can be stored as objects with very low retrieval times. Individual S3 objects can be as large as 5 TB.

We will use US-East-2 as our default AWS region since this is the cheapest among all AWS geographical regions, but in general you should pick a geographical region closest to where your data and computing resources are to minimize latency. Make sure that you create a new IAM user which has the IAM policy called "AmazonS3FullAccess" before proceeding with this section. When creating a new user, you should enable programmatic access and management console access so that you can access S3 through web UI as well as through SDK discussed in the next section.

The standard S3 storage charge for the US-East-2 region is $23/TB/month, with $0.005/1000 requests for accessing the data (via PUT, COPY, POST, LIST requests). Standard S3 does not charge a per GB fee for retrieving the data, and there is no minimum storage duration so you will be only charged for the actual storage time of an object. It may seem that this is extremely expensive compared to consumer storages such as Dropbox, but their business model and user applications are very different; with Dropbox, you get charged a fixed monthly fee for a data allowance which you may not use fully, and indeed most customers don't use anywhere near their top end limit. In case you do use the top end limit of Dropbox, then it will indeed be cheaper than Dropbox. However, most enterprises use S3 as their general-purpose object store, data lake, and so on, and with S3 you only pay for data you actually store.

S3 also offers a long-term, low-cost backup solution called Glacier suitable for long-term storage where the retrieval times are between 3 and 5 hours. These cost about $4/TB/month but have a standard retrieval charge of $10/TB and $0.05/1000 retrieval requests. There is a minimum of 90-day storage charge with S3 Glacier. An even cheaper tier is Glacier Deep Archive with retrieval times in the range of 12–48 hours, and it costs about $0.99/TB/month and a bulk retrieval charge of $2.5/TB/month. There is a minimum of 180 storage charge with Glacier Deep Archive so this is cost-effective only if you are going to access rarely (1–2 times a year). Glacier and Glacier Deep Archive are excellent archival tools, and we use them as our backup store, but they really are unsuitable for a lot of applications due to minimum storage time requirements, and hence for the rest of the book, we will only use the S3 standard storage whenever we mention storing objects in S3.

If you are uploading objects into S3 using the AWS Management Console, it prompts you to select a storage class; please make sure you only select "Standard" and not any other options such as Glacier, Glacier Deep Archive, or Intelligent-Tiering since those are pretty inappropriate for our use case.

As mentioned earlier, the files stored within each S3 bucket are referred to as "objects," and the object names are called "keys" in the S3 world. So if you want to upload an image called image1.png to a bucket called website-data, then in S3 lingo we will say that you are storing an object with key image1. You can use slashes "/" in keynames, and these act like folders in an S3 bucket; so, for example, we could have stored image1 as images/image1, and this would be analogous to storing an image1 object in a folder called images.

S3 is eventually consistent for overwrite on an object, meaning that there is a chance that you get an older version of an object if it's been updated recently.

Creating a bucket

Please go to the S3 console (`https://s3.console.aws.amazon.com/`) and click the "create bucket" button. Select an appropriate bucket name, keeping in mind that AWS doesn't allow uppercase letters or underscores and bucket names must be unique among all existing bucket names in S3. See Figure 3-14. As a default, all public access is also being blocked, and this should only be changed for a minority of use cases such as when hosting a static website on S3.

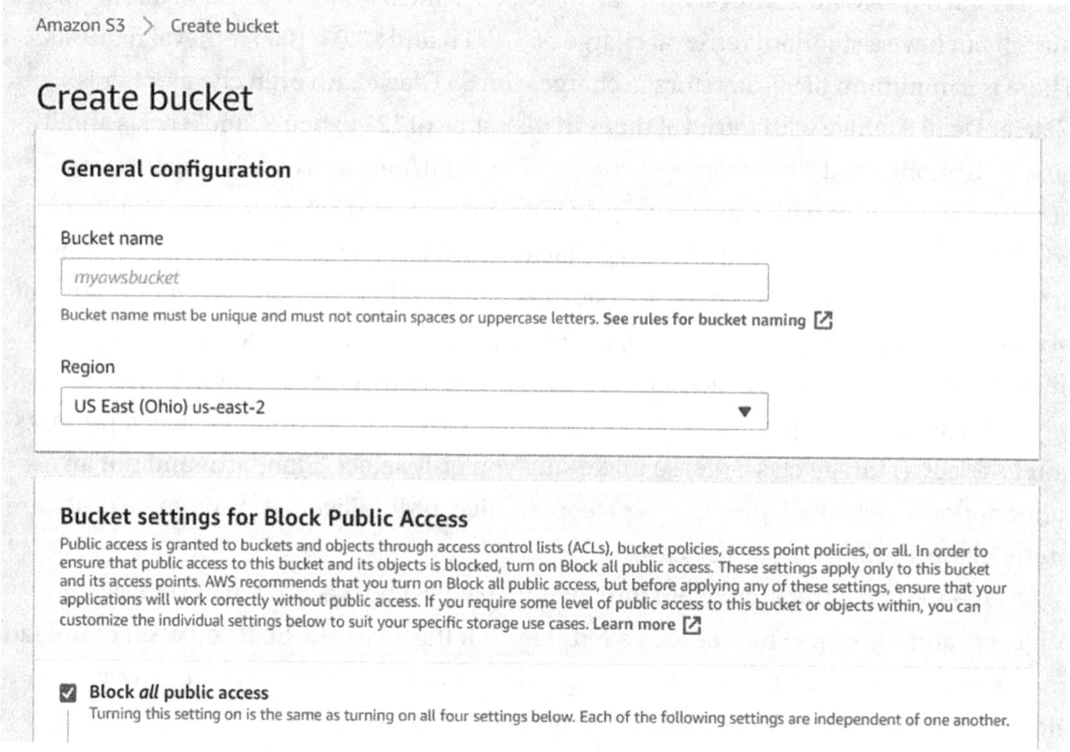

Amazon S3 > Create bucket

Create bucket

General configuration

Bucket name

myawsbucket

Bucket name must be unique and must not contain spaces or uppercase letters. **See rules for bucket naming**

Region

US East (Ohio) us-east-2

Bucket settings for Block Public Access

Public access is granted to buckets and objects through access control lists (ACLs), bucket policies, access point policies, or all. In order to ensure that public access to this bucket and its objects is blocked, turn on Block all public access. These settings apply only to this bucket and its access points. AWS recommends that you turn on Block all public access, but before applying any of these settings, ensure that your applications will work correctly without public access. If you require some level of public access to this bucket or objects within, you can customize the individual settings below to suit your specific storage use cases. **Learn more**

☑ **Block *all* public access**
   Turning this setting on is the same as turning on all four settings below. Each of the following settings are independent of one another.

Figure 3-14. *Create an S3 bucket*

Once a new bucket is created, you can upload the files through their web UI and select the storage class as shown in Figure 3-15.

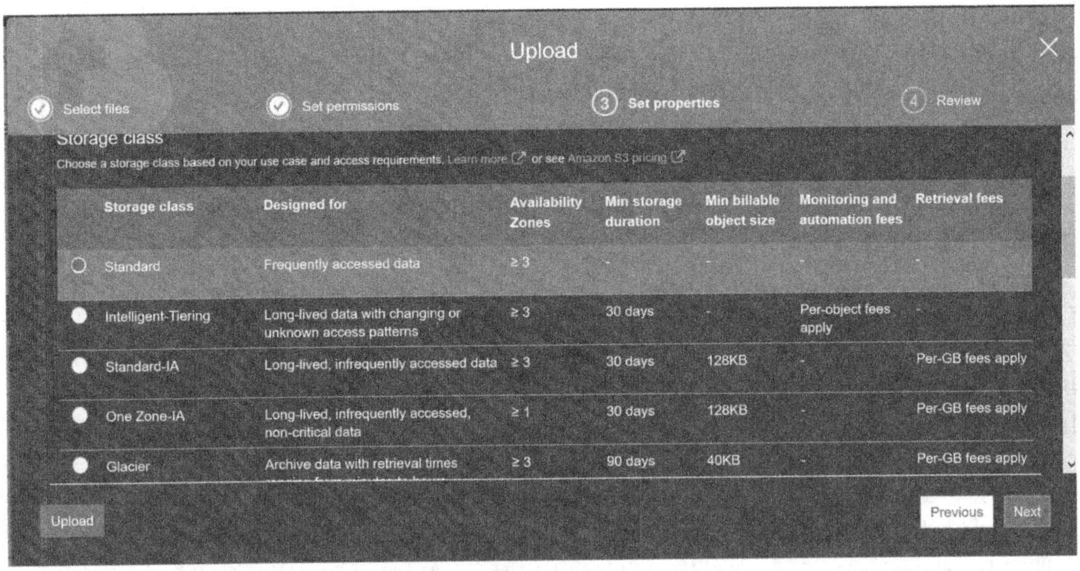

Figure 3-15.   Create an S3 bucket (cont.)

Accessing S3 through SDKs

I think it's time we switch to the Python SDK library for AWS called Boto3 which will allow us to perform S3 operations programmatically.

Boto3 has an object called "resource" which provides a high-level abstraction to a low-level interface called "clients" which almost mirrors the AWS service APIs.

You can pass IAM authentication details such as aws_access_key_id and aws_secret_access_key as parameters in boto3 client calls, but this is NOT the recommended way to do it due to obvious security issues. These details are visible to you on the IAM dashboard.

You can save these as environmental variables, but I think the best method is to download and install AWS Command Line Interface (CLI) and call aws configure to set up ~/.aws/config and ~/.aws/credentials files.

```
$ aws configure

AWS Access Key ID [None]: Enter_AWS_KEY_ID
AWS Secret Access Key [None]: Enter_secret_access_key
Default region name [None]: Enter default AWS region
```

Alternately, you can set up the credentials file yourself; it's just a text file with the information as follows:

101

```
[default]
aws_access_key_id = YOUR_ACCESS_KEY
aws_secret_access_key = YOUR_SECRET_KEY
```

For Windows users, the .aws folder will be in the C:\Users\user_name folder.

Let us perform the same set of operations starting with creating a new bucket as shown in Listing 3-2.

Listing 3-2. Creating an S3 bucket

```
import logging
import boto3
from botocore.exceptions import ClientError

def create_bucket(bucket_name, region, ACL_type):
    '''pick an ACL from 'private'|'public-read'|'public-read-
write'|'authenticated-read' '''

    # Create bucket
    try:

        s3_client = boto3.client('s3', region_name=region)
        location = {'LocationConstraint': region}
        s3_client.create_bucket(ACL = ACL_type, Bucket=bucket_name,
                                CreateBucketConfiguration=location)
    except ClientError as e:
        print(str(e))
        return False
    return True
create_bucket("test-jmp-book", 'us-east-2', 'private')
#Output
True
```

Once we have a new bucket, let's upload a file called test.pdf to it as shown in Listing 3-3. The maximum file size for each PUT call is 5 GB; beyond that, you will need to use multipart upload; it is frequently recommended that a multipart upload be used for files exceeding 100 MB.

Listing 3-3. Uploading an object to S3

```
import boto3
def S3_upload(S3_bucket_name,local_filename, S3_keyname):
    s3 = boto3.client('s3')

    for attempt in range(1,6):
        try:
            # files automatically and upload parts in parallel.
            s3.upload_file(local_filename,S3_bucket_name, S3_keyname)

        except Exception as e:
            print(str(e))
        else:
            print("finished uploading to s3 in attempt ", attempt)
            break
S3_upload("test-jmp-book", "test.pdf", "upload_test.pdf")
#output
finished uploading to s3 in attempt  1
```

We have used five retries in the preceding code just to make the code a bit robust against network issues.

Now, let us do something a bit more complicated; let's upload two to three more objects to the S3 bucket using the same code, and then let us try to query the last modified file as shown in Listing 3-4. We have included an option to filter the results by file extensions as well as by substring searching so that we can apply this function elsewhere.

Listing 3-4. Get the last modified file from a bucket

```
from datetime import datetime

def get_last_mod_file(s3bucketname, file_type = None, substring_to_match = ''):

    s3 = boto3.resource('s3')

    my_bucket = s3.Bucket(s3bucketname)
```

```
    last_modified_date = datetime(1939, 9, 1).replace(tzinfo=None)
    if any(my_bucket.objects.all()) is False:
        last_modified_file = 'None'
    for file in my_bucket.objects.all():
    # print(file.key)

        file_date = file.last_modified.replace(tzinfo=None)
        file_name = file.key
        print(file_date, file.key)
        if file_type is None:
            if last_modified_date < file_date and substring_to_match in
            file_name:
                last_modified_date = file_date
                last_modified_file = file_name
        else:

            if last_modified_date < file_date and substring_to_match in
            file_name and file_type == file_name.split('.')[-1]:
                last_modified_date = file_date
                last_modified_file = file_name
    return(last_modified_file)
get_last_mod_file("test-jmp-book")
```

Downloading files from the S3 bucket is pretty simple too as shown in Listing 3-5, and the approach mirrors our upload function. One thing to note is that we can download a file and save it as a name different than the keyname present in the S3 bucket.

Listing 3-5. Downloading an object from S3

```
def download_file_from_s3(s3bucketname,S3_keyname,local_filename):

    s3 = boto3.resource('s3')

    for attempt in range(1,6):

        try:

            s3.meta.client.download_file(s3bucketname, S3_keyname, local_
            filename)
```

```
    except botocore.exceptions.ClientError as e:
        if e.response['Error']['Code'] == "404":
            print("The object does not exist.")

    except Exception as e:
        print(e)
        logging.info(str(e))
    else:
        print("downloaded successfully in attempt ", attempt)

        break
download_file_from_s3("s3-jmp-upload-test","upload_test.pdf", "download_
test.pdf")
#output
downloaded successfully in attempt  1
```

Deleting an object from a bucket is straightforward; however, deleting all objects does require us to iterate through all objects and save them by their keyname and version id. You can only delete a bucket if it contains no object so it's very important to bulk delete all objects before trying to delete a bucket as shown in Listing 3-6.

Listing 3-6. Deleting an object from a bucket and the bucket itself

```
S3 = boto3.client('s3')
bucket_name = 'test-jmp-book'
key_name = 'upload_test.pdf'

response = S3.delete_object(
    Bucket=bucket_name,
    Key=key_name,
)

# deleting a bucket

def delete_all_objects(bucket_name):
    result = []
    s3 = boto3.resource('s3')
    bucket=s3.Bucket(bucket_name)
    for obj_version in bucket.object_versions.all():
```

```
            result.append({'Key': obj_version.object_key,
                           'VersionId': obj_version.id})
    print(result)
    bucket.delete_objects(Delete={'Objects': result})

def delete_bucket(bucket_name):
    s3 = boto3.resource('s3')
    my_bucket = s3.Bucket(bucket_name)
    if any(my_bucket.objects.all()) is True:
        delete_all_objects(bucket_name)
    my_bucket.delete()
    return True
delete_bucket('test-jmp-book')

# output

[{'Key': 'upload_test.pdf', 'VersionId': 'null'}]
True
```

Lastly, we can confirm if the bucket is indeed deleted by querying for all buckets as shown in Listing 3-7.

Listing 3-7. List all buckets

```
# Retrieve the list of existing buckets

def list_buckets():
    s3 = boto3.client('s3')
    response = s3.list_buckets()

    for bucket in response['Buckets']:
        print({bucket["Name"]})
        print('*'*10)
list_buckets()
```

Cloud storage browser

Even though S3 contains a web UI and an excellent SDK, it's still not as convenient as Dropbox, Google Drive, or Azure onedrive when you are trying to browse through your buckets looking for a specific file. To bridge the gap, a lot of companies that rely on S3 have rolled their own solution within their web apps or intranet sites in part since it's really difficult to upload large files (>100 MB) using the web UI without getting errors (even though the official limit is 5 GB for a web UI).

Let's talk about a general-purpose client called Cyberduck which gives you the best of both worlds, a UI as well as the ability to upload large files which makes it perfectly complementary to boto3 and a substitute for the web UI.

Cyberduck is a free software application available for download here for Mac and Windows. Once you have downloaded it, open it and click open connections on the Cyberduck dashboard as shown in Figure 4-16.

Figure 3-16. *Cyberduck dashboard*

Select S3 from the drop-down menu and enter the access key id and secret access key of your IAM user as shown in Figure 3-17.

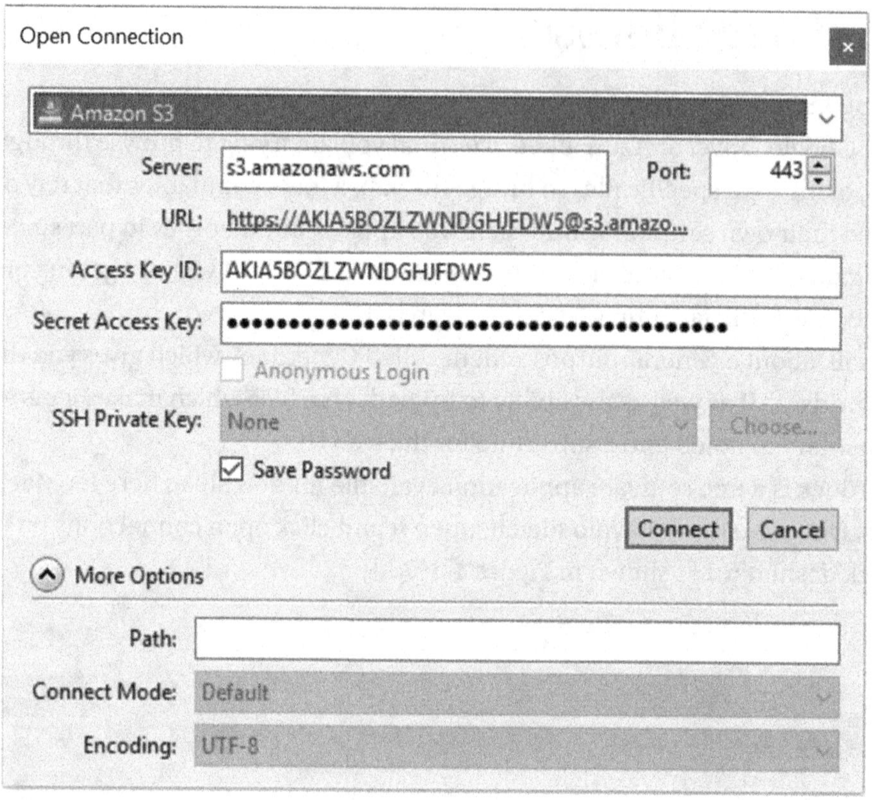

Figure 3-17. *Cyberduck connections*

In case you want to open a specific bucket, then simply enter the bucket name in the path text box earlier. You can leave it blank in case you want to go to the root page which lists all buckets. Creating a new bucket is simple too; right-click the root page, select new folder, enter the bucket name, and select region in the prompt shown in Figure 3-18.

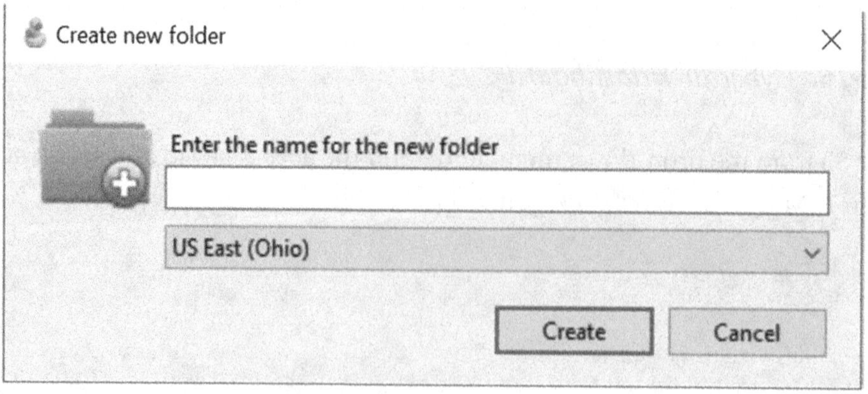

Figure 3-18. *Setting a bucket name and region*

Lastly, advanced bucket settings are also accessible from the root page; highlight the bucket name and click the Get info button on the top pane; that should give you a pop-up page shown in Figure 3-19 which includes check boxes to activate advanced settings such as object versioning or transitioning objects to S3 Glacier and so on, all of which are deactivated by default.

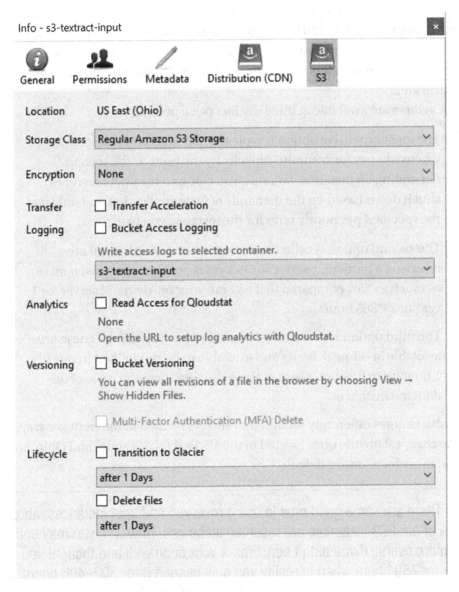

Figure 3-19. *Cyberduck S3 options*

Amazon EC2

Amazon Elastic Compute Cloud (Amazon EC2) is a web service that provides scalable compute capacity in the cloud. Basically, you can start and stop a server in a geographical area of your choice with desired configurations and can either pay by the second or can reserve it for a longer duration.

From a technical standpoint, EC2 has a wide variety of servers, from high RAM instances to high CPU cores, and in the past couple of years, they also have GPU instances which are predominantly used for training deep learning models.

An AWS EC2 instance refers to a particular server operated from a specific geographical area.

All the servers are available at three distinct price points:

- The most expensive option is typically "On-Demand instances," where you pay for compute capacity by the hour or the second depending on instances you run. You can spin up a new server or shut it down based on the demands of your application and only pay the specified per hourly rates for the instance you use.

- The second option is called "reserved instances"; here you are reserving a particular server for 1–3 years, and you get a discount of as much as 75% compared to if you ran your on-demand server for 1 year or 24*365 hours.

- The third option is something known as "spot instances." These are available in off-peak hours and available to be run in increments of 1 hour up to 6 hours. There is also a discount over the price of on-demand instances.

AWS also charges differently for different servers located in different geographical areas. The cheapest are the ones located in the US East (N. Virginia) and Ohio, and we always use those locations as default.

Note   There are some good pros in using reserved and spot instances; about 65–70% of our EC2 instances are reserved instances. However, you may end up paying more overall if you didn't benchmark your code well and thought you need a server for 750 hours when in reality you only needed it for 300–400 hours, and the rest of the hours were wasted. In that scenario, you'll save money by paying on-demand pricing and shut off the server when work is complete.

EC2 server types

EC2 has very tiny servers for small loads called t2.nano ones which only cost $0.0058 per hour (or $5/month) and has one CPU core and 512 MB RAM; there are other general-purpose servers in this category which are more powerful than nano which can host high traffic websites, intensive crawling tasks, and so on.

There are high CPU/compute servers such as "c5n.18xlarge" which costs about $3.9/hour (~$2900/month) and has 72 cores and 192 GB RAM. We mainly use these high compute servers for training CPU-intensive machine learning models.

For training deep learning models in TensorFlow/PyTorch, there are also GPU instances available such as p3.16xlarge which costs about $24/hour (~$18,000/month) and has Nvidia GPUs for fast computations.

If in doubt, you should use the smallest compute-optimized server available, which is c5.large ($0.085/hr or $63/month) and go up as needed. At Specrom Analytics, we get all our analysts started out with c5.large before going for higher configurations for specific jobs.

In addition to paying compute costs for EC2, you also have to pay data transfer costs which are about $90/TB for data transfer out to the Internet and about $10–20/TB for data transfer to other AWS products such as S3 depending on the region. Now, you can minimize this cost of data transfer to free for S3, DynamoDB, and so on if they are in the same region as your EC2 instance and to a flat rate of $10/TB for VPC peering with Amazon RDS.

EC2 also needs a local storage akin to a hard drive, and this is called "Elastic Block Storage (EBS)"; these cost about $30–$135/TB/month depending on the SSD of the EBS storage type. EBS Snapshots are charged at $50/TB/month.

Lastly, if you want a fixed IP address for your EC2 instance such as when hosting a website, then you can get one free elastic IP address linked to your running EC2 instance at no charge. However, you will pay $0.005/hour for the duration that your elastic IP address is not being associated with the EC2 instance, such as when the instance is shut down.

We will stick to web UI for creating new EC2 instances so that there is less chance of an inadvertent error since provisioning the wrong instance type could result in getting a credit card bill of thousands of dollars at the end of the month.

Spinning your first EC2 server

Log in to your AWS and go to the EC2 dashboard as shown in Figure 3-20. Click the "Launch instance" button. You should shut off the servers when they are not being used to stop incurring charges.

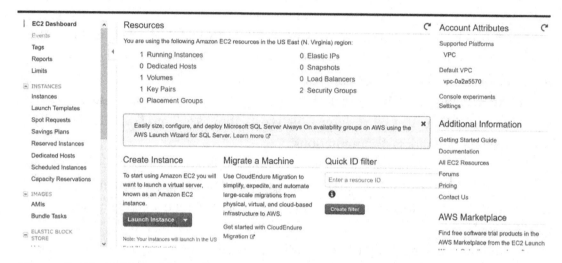

Figure 3-20. *EC2 dashboard*

You can choose the plain vanilla server with only OS installed; but in actual development, we tend to spin servers from Amazon Machine Images (AMI) as shown in Figure 3-21 which already come with software packages installed for intended tasks. For example, if you want to run a WordPress website from an EC2 instance, then you can get an AWS Marketplace which includes a MySQL, Apache web server, and WordPress so that you have the entire stack ready for web hosting.

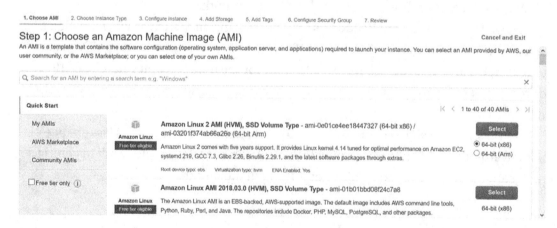

Figure 3-21. *AMI images on the AWS Marketplace*

Let us search for AMIs with Anaconda preinstalled shown in Figure 3-22 so that we can quickly start using our EC2 instance. Once you have created your EC2 server, I recommend that you create an AMI image for it so that you can spin multiple servers with the same environment with minimal effort.

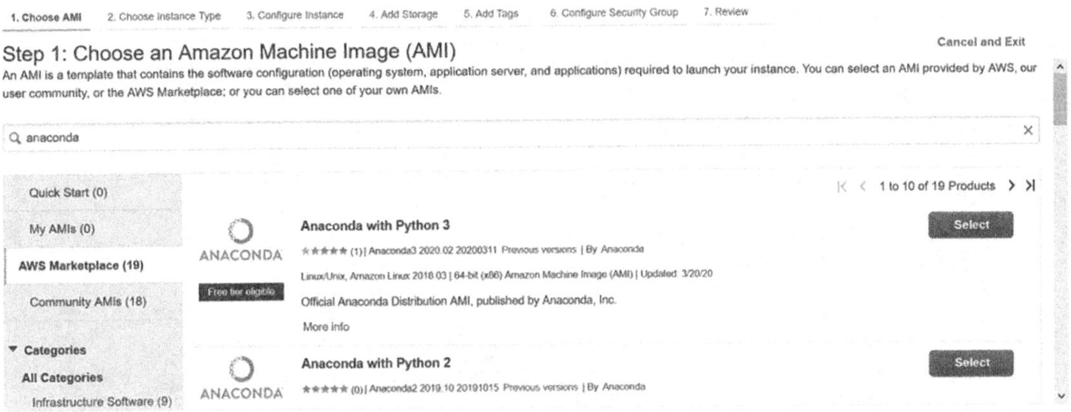

Figure 3-22. *AMI with an Anaconda package*

Some of the AWS Marketplace AMIs have a subscription charge in addition to EC2 server costs, but for most open source software packages, you should be able to find something for free. You can see these additional charges under "software"; as you can see, it's zero for Anaconda shown in Figure 3-23.

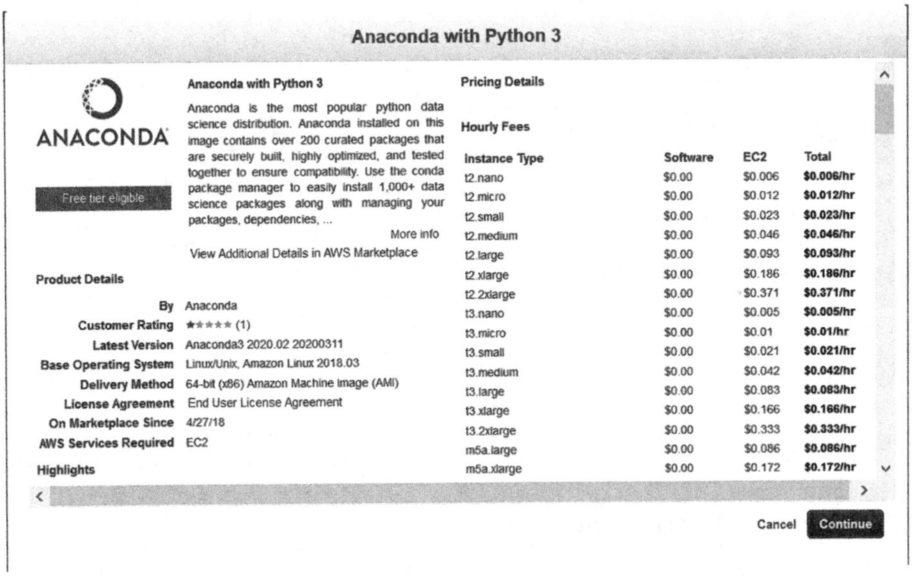

Figure 3-23. *More information about Anaconda-based AMI*

Once you click continue, you can select the instance type shown in Figure 3-24; the recommended instance (c5.large) is NOT eligible for free tier. If you want, pick t2.micro which is free tier eligible; however, make sure that you spin another EC2 instance C5.large or better for running computationally intensive tasks mentioned in other chapters.

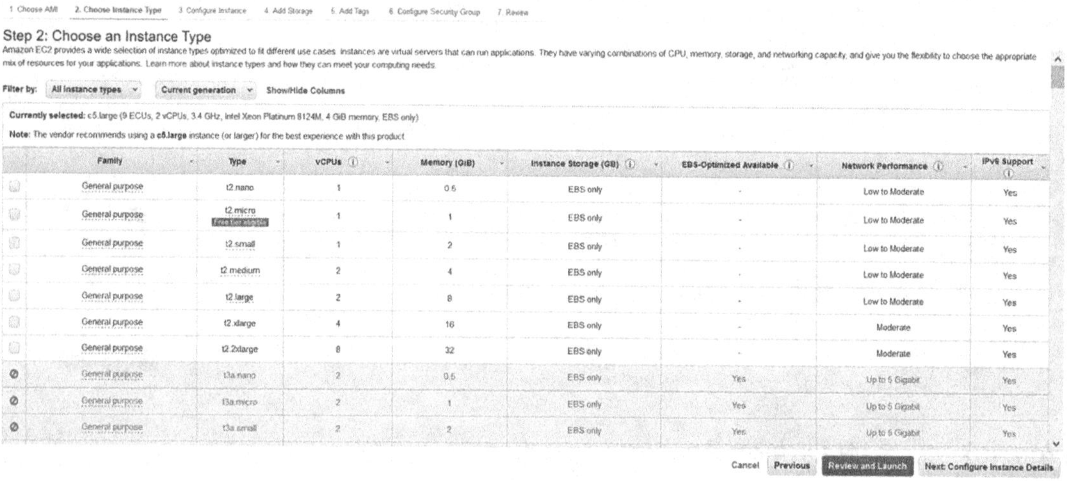

Figure 3-24. *EC2 server types*

Continue next without changing anything on the configure instance step and edit the storage to 30 GB from the default 8 GB and add tags shown in Figure 3-25.

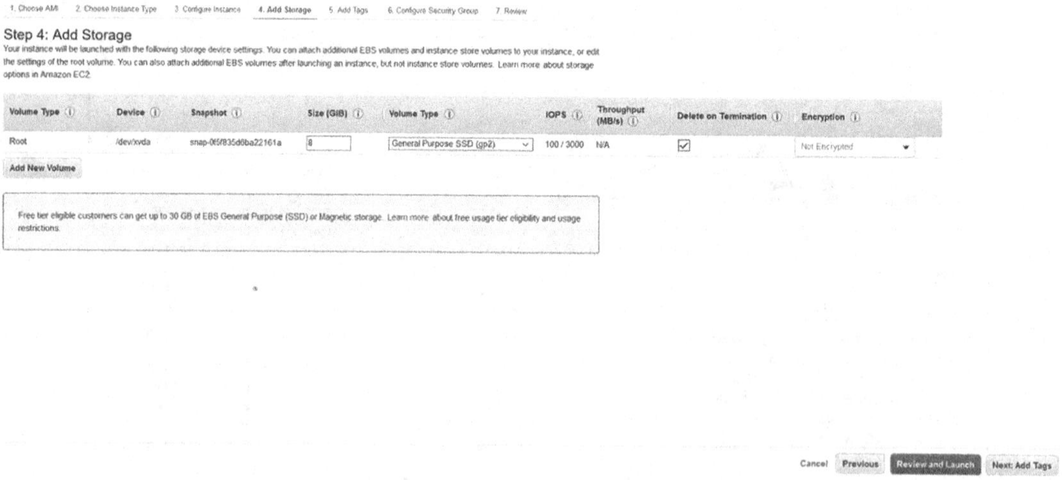

Figure 3-25. *EBS storage option*

Click next for steps 5 and 6 without any modifications, and finally when you click launch on step 7 shown in Figure 3-26, it should tell you to create a key pair. This step is important; create one key and save it on your computer. You will not be able to communicate with your server without this key.

Figure 3-26. *Final EC2 server launch details*

Now go to the AWS dashboard from the left corner, and pick the ec2 dashboard; click the running instance link and go to more details. Save the public DNS (IPv4); you will need this to communicate with the server using SSH.

You can assign an IAM role to this EC2 server by clicking Actions ➤ Instance Settings ➤ Attach/Replace IAM Role as shown in Figure 3-27.

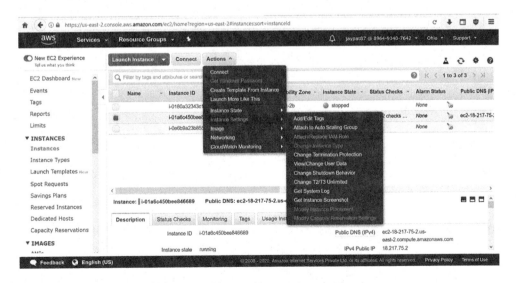

Figure 3-27. *EC2 server settings from the dashboard*

Communicating with your EC2 server using SSH

Let's communicate with your EC2 server instance from your local computer using Secure Shell (SSH) protocol. Linux distributions already come prepackaged with openSSH, and all you need to do is change the permissions of your .pem file to 400 by running

```
chmod 400 my-key-pair.pem
```

Enter the following command to initiate a session with your EC2 server. Note that ec2-user is the default username of EC2 instances with Linux distributions such as the one we have on our first server instance, and my-key-pair.pem is the key file you downloaded when creating a new ec2 instance.

```
ssh -i /path/my-key-pair.pem ec2-user@public_DNS(IPv4)_address
```

If you are using Windows on a local computer, then you will have to download an SSH client called PuTTY (`www.putty.org/`).

PuTTY only accepts keys in .ppk format, and unfortunately AWS uses a .pem key. So we will have to download PuTTYgen (`www.puttygen.com/`) to convert the key formats.

Click load an existing private key and select the path to your .pem file. Once it's loaded, click save private key and a .ppk file will be downloaded (see Figure 3-28). Use this ppk file in PuTTY in the next step.

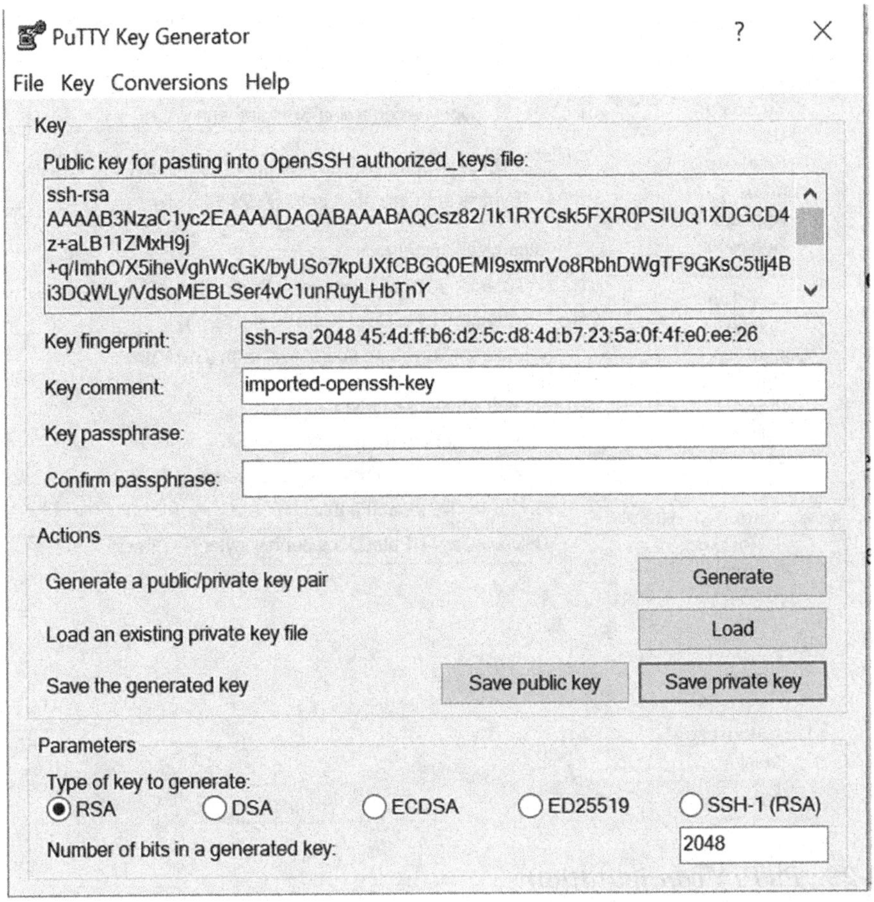

Figure 3-28. *PuTTY key generator*

Next, go to SSH and click Auth and browse to the location of the .ppk key file (Figure 3-29).

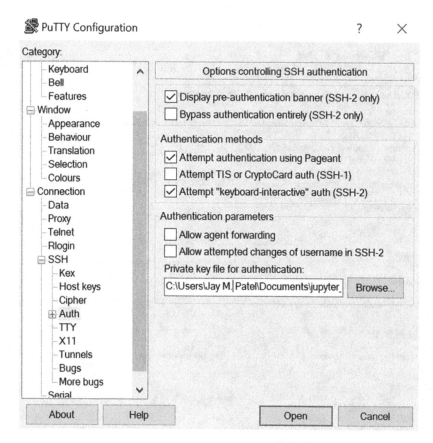

Figure 3-29. *PuTTY configuration*

Copy the public DNS IP (IPv4) address you got from EC2 dashboard to the Host name field in PuTTY. Don't forget to append the username which is "EC2-user" followed by @ before the IP address as shown in Figure 3-30. Click open to get connected to your EC2 server.

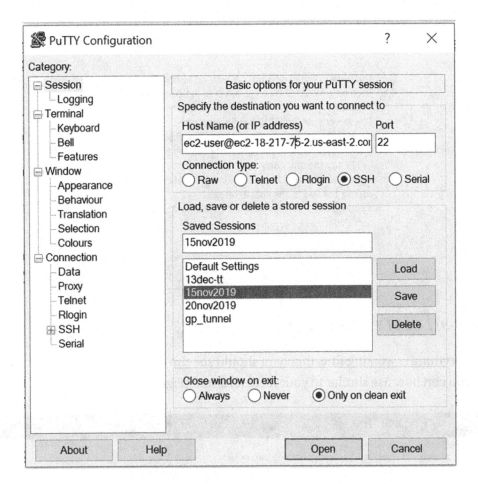

Figure 3-30. *PuTTY configuration (cont.)*

You will get a security warning as shown in Figure 3-31; just click yes.

Figure 3-31. *PuTTY security alert*

If everything is working fine, then you should see the terminal window open up which you can now use similar to your local terminal (Figure 3-32).

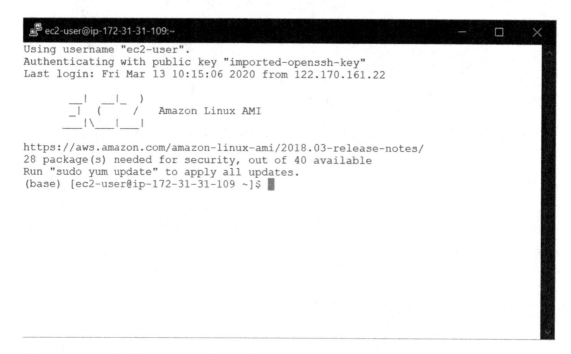

Figure 3-32. *PuTTY terminal window*

You can now start to move your Python scripts from local to a remote server using SFTP as discussed in the next section. Note that you already have Anaconda distribution installed on your machine, so running your Chapter 2 scripts shouldn't take long at all.

You can verify your preinstalled Anaconda version and other package versions by calling

```
conda list
```

See Figure 3-33.

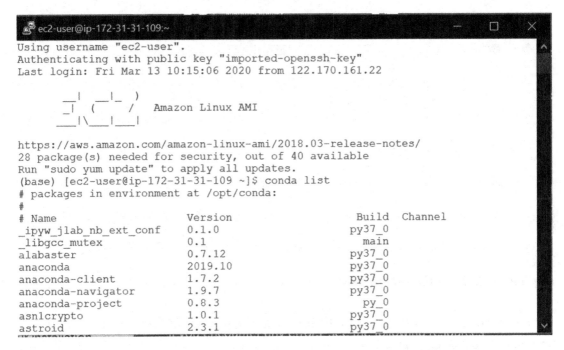

Figure 3-33. *List of Anaconda packages on the PuTTY terminal*

Transferring files using SFTP

Now that you have a terminal through SSH, I am sure you are wondering how to run transfer files to the remote server. We can use simple file transfer protocol (SFTP) clients to upload/download files from a remote server to local.

We already saw an SFTP client called Cyberduck in the S3 section, and you can continue to use it for transferring files to EC2. However, there are some reports on forums regarding intermittent connection issues with transferring files to/from EC2; so let us briefly cover another popular SFTP client called FileZilla which is available on Mac, Windows, and Linux.

Open FileZilla and click File on the top-left corner and select site manager (Figure 3-34).

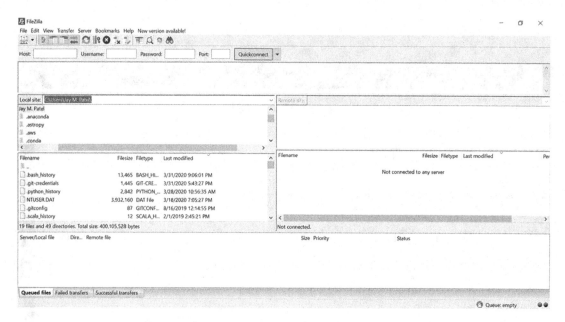

Figure 3-34. *FileZilla dashboard*

Enter the public DNS address in the host text box, username, and key file location and hit connect. See Figure 3-35.

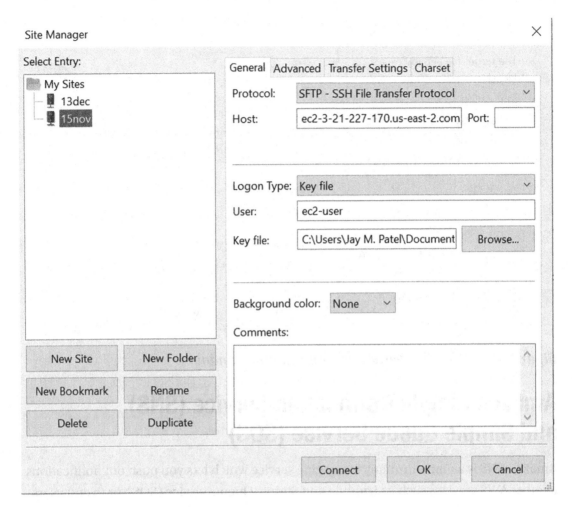

Figure 3-35. *FileZilla settings*

Once you are connected, you should be able to see the remote server file explorer on the right pane and the local computer file explorer on the left. You can transfer files between them by simply double-clicking a file of your choice. In case you are trying to download/upload a file with the same name, then you will see a notice shown in Figure 3-36, and you can select an appropriate action.

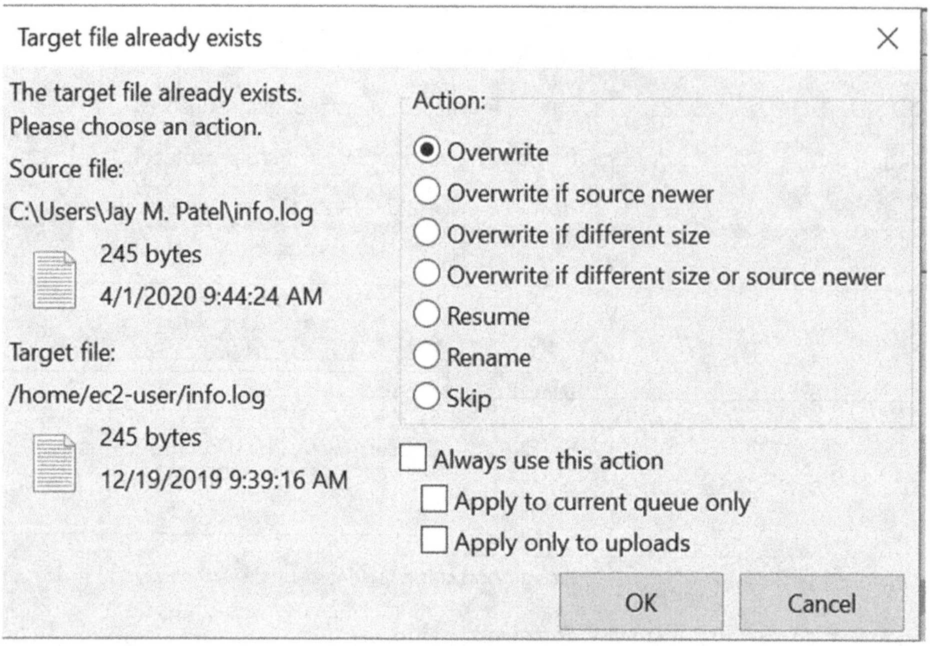

Figure 3-36. *FileZilla overwriting confirmation window*

Amazon Simple Notification Service (SNS) and Simple Queue Service (SQS)

Amazon SNS is a many-to-many messaging service which lets you push out notifications to other AWS services such as Lambda and Simple Queue Service (SQS) or to end users using mobile SMS, email, and so on.

SQS is a queue service which lets you build complete microservices; they have a standard queue which only has a best effort ordering and a FIFO queue which preserves the order the messages are sent. Unfortunately, FIFO SQS queues are not compatible with SNS so we will only stick to the standard queue here.

You may be wondering why we are bothering with setting up SQS and SNS. That's a fair question, but rest assured it represents an important component of cloud computing and can be used to pass messages back and forth between different cloud components, allowing us to decouple individual data processing pipelines and easily perform distributed batch processing in later chapters.

A simple use case for SNS and SQS is when you want to send a notification to a software developer via email to indicate that a particular batch processing job on EC2 has been finished. You can simultaneously also send a message to SQS which can be used as a trigger by a different AWS Lambda service.

Please attach the "AmazonSNSFullAccess" and "AmazonSQSFullAccess" policies to the IAM user before proceeding with this section.

The first step in setting up SNS is creating a new topic; this is where the publisher code will push messages which can then be forwarded by SNS to its subscribers; as discussed earlier, one SNS topic can have multiple subscribers so that everyone gets the message at the same time.

Let us create a new SNS topic; set up an email and SQS as a subscriber using boto3 library as shown in Listing 3-8. We have set the IAM policy which allowed SQS to access SNS through the set_queue_attributes(). This method is especially useful when we only plan to use a particular SNS/SQS for a particular task and delete it after our work is over. We have also made extensive use of Amazon Resource Names (ARNs) which uniquely identify AWS resources and are indispensable for programmatically accessing AWS services.

Listing 3-8. Creating SNS topic and SQS queue

```python
import boto3
import json
import sys
import time

def CreateTopicandQueue(topic_name, email_address):

        sqs = boto3.client('sqs')
        sns = boto3.client('sns')
        millis = str(int(round(time.time() * 1000)))

        #Create SNS topic
        snsTopicName=topic_name + millis

        topic_response=sns.create_topic(Name=snsTopicName)
        snsTopicArn = topic_response['TopicArn']

        # subscribing email_address to SNS topic
```

```
    if email_address is not None:

        email_response = sns.subscribe(TopicArn=snsTopicArn,Protocol='e
        mail',Endpoint=email_address,
    ReturnSubscriptionArn=True)
        emailArn = email_response['SubscriptionArn']

    else:
        emailArn = None

    #create SQS queue
    sqsQueueName=topic_name + millis
    sqs.create_queue(QueueName=sqsQueueName)
    sqsQueueUrl = sqs.get_queue_url(QueueName=sqsQueueName)['QueueUrl']

    attribs = sqs.get_queue_attributes(QueueUrl=sqsQueueUrl,

    AttributeNames=['QueueArn'])['Attributes']

    sqsQueueArn = attribs['QueueArn']

    # Subscribe SQS queue to SNS topic
    sns.subscribe(
        TopicArn=snsTopicArn,
        Protocol='sqs',
        Endpoint=sqsQueueArn)

    #Authorize SNS to write SQS queue
    policy = """{{
"Version":"2012-10-17",
"Statement":[
  {{
    "Sid":"MyPolicy",
    "Effect":"Allow",
    "Principal" : {{"AWS" : "*"}},
    "Action":"SQS:SendMessage",
    "Resource": "{}",
    "Condition":{{
```

```
        "ArnEquals":{{
          "aws:SourceArn": "{}"
        }}
      }}
    }}
  ]
}}""".format(sqsQueueArn, snsTopicArn)

        response = sqs.set_queue_attributes(
            QueueUrl = sqsQueueUrl,
            Attributes = {
                'Policy' : policy
            })

        return({"snsTopicArn":snsTopicArn,"sqsQueueArn":sqsQueueArn,"sqsQue
        ueUrl":sqsQueueUrl, 'emailArn':emailArn })

response_dict = CreateTopicandQueue("test_topic", "your_email_address")

# Output:
{'emailArn': 'arn:aws:sns:us-east-2:896493407642:test_
    topic1585456350589:273d413a-5484-4a96-a167-43791f45266f',
    'snsTopicArn': 'arn:aws:sns:us-east-2:896493407642:test_
    topic1585456350589',
    'sqsQueueArn': 'arn:aws:sqs:us-east-2:896493407642:test_
    topic1585456350589',
    'sqsQueueUrl': 'https://us-east-2.queue.amazonaws.com/896493407642/test_
    topic1585456350589'}
```

Even though we are getting email ARN from sns.subscribe(), you should know that it becomes active only if the subscriber clicks the confirmation email sent out by AWS. Sending a message to SNS is pretty simple; you just need to specify topicARN as shown in Listing 3-9.

Listing 3-9. Sending a message through SNS

```
client = boto3.client('sns')
response = client.publish(
    TopicArn=response_dict["snsTopicArn"],
```

```
    Message='this is a test of SNS and SQS',
    Subject='test_SNS_SQS',
    MessageStructure='string',
)
```

Retrieving messages from the SQS queue is a bit more involved as shown in Listing 3-10. You can get a maximum of ten messages per request, and long polling can be enabled by setting a longer time in seconds for the WaitTimeSeconds parameter. sqsResponse consists of the Messages key and ResponseMetadata; you will not get a Messages key if your queue is empty. Deleting a message requires that you specify receipthandle.

Listing 3-10. Retrieving messages through SQS

```
sqs = boto3.client('sqs')
sqsResponse = sqs.receive_message(QueueUrl=response_dict['sqsQueueUrl'], Me
ssageAttributeNames=['ALL'], MaxNumberOfMessages=10, WaitTimeSeconds = 10)
# parsing sqs messages
sqsResponse["Messages"]

if 'Messages' in sqsResponse:
    for message in sqsResponse["Messages"]:
        message_dict = json.loads(message["Body"])
        message_text = message_dict["Message"]
        subject_text = message_dict["Subject"]
        message_id = message_dict["MessageId"]
        receipt_handle = message["ReceiptHandle"]
        #print("receipt_handle: ", receipt_handle)
        print("message_id: ", message_id)
        print("subject_text: ", subject_text)
        print("message_text: ", message_text)
# Output:
message_id:  45bc89a2-ab35-52f8-bd49-df5899a958c9
subject_text:  test_SNS_SQS
message_text:  this is a test of SNS and SQS
```

Deleting a message from a queue as well as deleting an SQS queue and SNS topic is simple enough as shown in Listing 3-11; however, note that unlike deleting an S3 bucket where AWS will not allow you to delete a bucket with objects, here there are no safeguards in place by AWS so it's perfectly valid to delete an SQS queue with messages still inside, so you will have to provide that safeguard in your code.

Listing 3-11. Deleting a message by receipt handle and deleting an SNS and SQS queue

```
response = sqs.delete_message(
    QueueUrl=response_dict['sqsQueueUrl'],
    ReceiptHandle=receipt_handle
)

# DELETE SQS and SNS queue

sqs.delete_queue(QueueUrl=response_dict['sqsQueueUrl'])
sns.delete_topic(TopicArn=response_dict['snsTopicArn'])
```

Scraping the US FDA warning letters database on cloud

We will migrate the script from Listing 2-27 from Chapter 2 on an EC2 server with minor modifications as shown in Listing 3-12.

Our script is modified to upload the created CSV file to the S3 bucket. We have also inserted a send message to SNS/SQS so that we can be notified by email whenever the job is complete.

We have hardcoded the response_dict we got while creating an SNS and SQS queue.

The other hardcoded components include the website URL to scrape from as well as the S3 bucket and filenames. This is done to keep the script as simple as possible, but I highly suggest that you never do it in production setting and instead load parameters from an external configuration JSON file.

We have also not created any error and information logging, and I think that's indispensable for anything but the most simplest scripts.

Listing 3-12. Sample script to run on the EC2 server

```python
#! /opt/conda/bin/python3

import requests
import numpy as np
import pandas as pd
import io
from bs4 import BeautifulSoup
import boto3

def S3_upload(S3_bucket_name,local_filename, S3_keyname):
    S3 = boto3.client('s3')

    for attempt in range(1,6):
        try:
            # files automatically and upload parts in parallel.
            S3.upload_file(local_filename,S3_bucket_name, S3_keyname)

        except Exception as e:
            print(str(e))
        else:
            print("finished uploading to s3 in attempt ", attempt)
            break

def get_abs_url(html_tag):
    soup = BeautifulSoup(html_tag,'lxml')
    abs_url = 'https://www.fda.gov' + soup.find('a')['href']
    company_name = soup.find('a').get_text()
    return abs_url, company_name

if __name__ == "__main__":  # confirms that the code is under main function

    my_headers = {
    'User-Agent': 'Mozilla/5.0 (Windows NT 10.0; Win64; x64)
    AppleWebKit/537.36 ' + ' (KHTML, like Gecko) Chrome/61.0.3163.100Safa
    ri/537.36'
    }
```

```
test_url = 'https://web.archive.org/save/_embed/https://www.fda.gov/
files/api/datatables/static/warning-letters.json?_=1586319220541'

response_dict = {'emailArn': 'arn:aws:sns:us-east-2:896493407642:test_
topic1586487525592:40239d22-7025-40b4-ac4b-bd36e3a1f9cc',
'snsTopicArn': 'arn:aws:sns:us-east-2:896493407642:test_
topic1586487525592',
'sqsQueueArn': 'arn:aws:sqs:us-east-2:896493407642:test_
topic1586487525592',
'sqsQueueUrl': 'https://us-east-2.queue.amazonaws.com/896493407642/
test_topic1586487525592'}

r = requests.get(url = test_url, headers = my_headers)
#print("request code: ", r.status_code)
html_response = r.text
string_json2 = io.StringIO(html_response)
df = pd.read_json(string_json2)
df["abs_url"], df["company_name"] = zip(*df["field_company_name_
warning_lette"].apply(get_abs_url))
df.to_csv("warning_letters_table.csv")
S3_keyname = "warning_letters_table.csv"
local_filename = "warning_letters_table.csv"
S3bucket_name = 'test-jmp-book'
S3_upload(S3bucket_name,local_filename,S3_keyname)
# sending a message through SNS
message_text = S3_keyname + " successfully uploaded to " + S3bucket_name
client = boto3.client('sns', 'us-east-2')
response = client.publish(
    TopicArn=response_dict["snsTopicArn"],
    Message=message_text,
    Subject='s3 upload successful',
    MessageStructure='string',
)
```

Simply upload the preceding script to the home directory of the EC2 server using FileZilla.

Connect to your EC2 server using PuTTY. Make the script an executable so that now it can be started directly by typing ./script.py instead of python script.py by typing

```
chmod a+x us_fda_script.py
```
Please confirm it by

```
ls -l
```
You should see the permissions listed as follows:

```
-rwxrwxr-x 1 ec2-user ec2-user    2105 Apr 10 03:25 us_fda_script.py
```
On the ec2 server, we will make the script an executable so that now it can be started directly by typing ./script.py instead of python script.py.

If you are writing your scripts on a Windows-based local computer, then you will have to remove the DOS-based line endings from the script before you can execute it on Linux.

```
sed -i -e 's/\r$//' us_fda_script.py
```
As a last step, we will use cronjobs to start the script automatically every time the server reboots. To open the crontab, enter

```
crontab -e
```
A VIM editor will open up; just type the command as follows and enter :wq to exit crontab:

```
@reboot /home/ec2-user/us_fda_script.py
```
Just confirm that you have correctly loaded it by typing

```
crontab -l
```
It should return back the location of the script.

```
@reboot /home/ec2-user/us_fda_script.py
```
At this point, we are ready to put our server to test. Go to the EC2 console, and click reboot on your server instance. If everything is in order, then you should get an email shortly which means that your script was triggered and it has uploaded the CSV file to the S3 bucket.

Congratulations! You just learned how to run your scripts on the cloud.

Summary

We have learned the basics of cloud computing in this chapter with an in-depth look at AWS's permissions (IAM), object storage (S2), computing servers (EC2), simple notification system, and simple queue system (SQS). We used these to run the script from the last chapter on the cloud.

In the next chapter, we will introduce natural language processing (NLP) techniques and their common applications in web scraping.

CHAPTER 4

Natural Language Processing (NLP) and Text Analytics

In the preceding chapters, we have solely relied on the structure of the HTML documents themselves to scrape information from them, and that is a powerful method to extract information.

However, for many use cases, that still doesn't get us specific enough information, and we have to use algorithms and techniques which work directly on raw text itself.

We will survey natural language processing (NLP) techniques and their common use cases in this chapter. The goal here is to present NLP methods and case studies illustrating their real-word application in the domain of web-scraped data.

I understand that many of my readers will not be familiar with machine learning in general or NLP in particular, and that's fine. I have tried to present the NLP material here as sort of a black box algorithm with minimal discussion on "how" the algorithm works and focusing solely on the problem at hand.

We will demonstrate applications of mainstream machine learning and NLP libraries such as sklearn, NLTK, Gensim, SpaCy, and so on and write glue code to make it work better or train the machine learning or deep learning/neural network model abstracted within the libraries. We will not show you how to mimic the functionalities contained in those libraries from scratch since that is outside the scope of this book.

If someone wants to learn about the fundamentals of NLP and information retrieval in general, they are directed to refer to the *Introduction to Information Retrieval* by

© Jay M. Patel 2020
J. M. Patel, *Getting Structured Data from the Internet*, https://doi.org/10.1007/978-1-4842-6576-5_4

Christopher D. Manning, Prabhakar Raghavan, and Hinrich Schütze.[1] Yoav Goldberg discusses the basics of neural networks for NLP in his paper,[2] and he expands this in his textbook.[3]

Similarly, we will not talk about word embeddings such as word2vec, GloVe, BERT, fastText, and so on or the most cutting-edge deep learning or neural network–based NLP strategies which can eke a few percent improvement accuracy gains over more mature and faster methods here, but you can check those out in the book.[4]

Regular expressions

Regular expressions (regex) match patterns with sequences of characters, and they are supported in a wide variety of programming languages. There is no learning happening in this case, and many argue that this should not even be part of any natural language processing chapter. That may very well be the case, but the fact remains that regex has been part of most software engineers' toolbox since the past three decades; and it's included as part of standard libraries in most programming languages including Python for that reason.

A common use case for regex is extracting or validating email addresses, datetimes, URLs, phone numbers, and so on from a text document. They are also widely used for search and replace in many commonly used programs and text processing software applications.

You should really use regex for a handful of well-defined and documented use cases. We are only going to mention an important regex use case from a web scraping perspective in this section before we move on to other NLP methods.

A complete tutorial on regex is outside the scope of this book, but I highly recommend going over an excellent introduction to regex article by Andrew Kuchling as part of the official Python documentation (`https://docs.python.org/3.6/howto/regex.html`).

---

[1]Cambridge University Press, 2008

[2]Goldberg, Yoav. "A primer on neural network models for natural language processing." *Journal of Artificial Intelligence Research* 57 (2016): 345-420)

[3]Goldberg, Yoav. "Neural network methods for natural language processing." *Synthesis Lectures on Human Language Technologies* 10.1 (2017): 1-309)

[4]Rao, Delip, and Brian McMahan. *Natural language processing with PyTorch: build intelligent language applications using deep learning.* O'Reilly Media, Inc., 2019

The *Regular Expressions Cookbook*, 2nd ed. by Jan Goyvaerts and Steven Levithan (O'Reilly, 2012) is also part of my reference collection, and I cannot praise it enough. Jan runs a very popular website called Regular-Expressions.info, so if you happen to google any regex-based questions, then chances are that you might have already benefited from Jan's pearls of wisdom.

I sometimes see inexperienced developers try to regex for all kinds of tasks such as HTML parsing which is an entirely unsuitable use case. One of the most linked answers by bobince on Stack Overflow (`https://stackoverflow.com/questions/1732348/regex-match-open-tags-except-xhtml-self-contained-tags/1732454#1732454`) pokes fun at this, but that still doesn't dissuade enough people to stop incorrect regex usage which not only makes your code harder to read and debug but far too fragile to put into production. When you first try to use a simple regex pattern to extract information from HTML, it appears like it's working except for a handful of special cases; so quite naively, you write a more complex regex to handle that, and at that point, you have forgotten about your original problem and started down the regex rabbithole.

I am sure this is what Jamie Zawinski refers to in his famous quote "Some people, when confronted with a problem, think 'I know, I'll use regular expressions.' Now they have two problems."

Extract email addresses using regex

We talked about the major use cases of web scraping in Chapter 1, and one of the websites I mentioned was Hunter.io which has built an email database by scraping a large portion of the visible Internet.

It allows you to enter a domain address, and it returns all the email addresses associated with that domain, as well as other useful meta-information such as the URL of the pages it first saw that email address, the dates, and so on.

Listing 4-1 shows results from their undocumented REST API which gives us all the information Hunter.io has about my personal website (jaympatel.com); note that you may get an error message back from the server since Hunter.io would probably be running antibot measures to prevent users from bypassing its front-end user interface and hitting the API endpoint directly.

Listing 4-1. Using Hunter.io for finding email addresses

```
import requests
import json
url = 'https://hunter.io/trial/v2/domain-search?limit=10&offset=0&domain=
jaympatel.com&format=json'
r = requests.get(url)
html_response = r.text
json.loads(html_response)
```

Output:

```
{'data': {'accept_all': False,
   'country': None,
   'disposable': False,
   'domain': 'jaympatel.com',
   'emails': [{'confidence': 94,
     'department': 'it',
     'first_name': 'Jay',
     'last_name': 'Patel',
     'linkedin': 'https://www.linkedin.com/in/jay-m-patel-engg',
     'phone_number': None,
     'position': 'Freelance Software Developer',
     'seniority': None,
     'sources': [{'domain': 'leanpub.com',
       'extracted_on': '2019-09-25',
       'last_seen_on': '2020-03-25',
       'still_on_page': True,
       'uri': 'http://leanpub.com/getting-structured-data-from-internet-web-
       scraping-and-rest-apis/email_author/new'},
      {'domain': 'leanpub.com',
       'extracted_on': '2019-09-04',
       'last_seen_on': '2020-05-11',
       'still_on_page': True,
       'uri': 'http://leanpub.com/u/jaympatel'},
```

```
    {'domain': 'jaympatel.com',
     'extracted_on': '2019-04-24',
     'last_seen_on': '2020-04-01',
     'still_on_page': True,
     'uri': 'http://jaympatel.com/cv'},
    {'domain': 'jaympatel.com',
     'extracted_on': '2019-04-22',
     'last_seen_on': '2020-04-01',
     'still_on_page': True,
     'uri': 'http://jaympatel.com/consulting-services'},
    {'domain': 'leanpub.com',
     'extracted_on': '2019-04-13',
     'last_seen_on': '2020-03-08',
     'still_on_page': True,
     'uri': 'http://leanpub.com/getting-structured-data-from-internet-web-
     scraping-and-rest-apis'},
    {'domain': 'jaympatel.com',
     'extracted_on': '2019-04-11',
     'last_seen_on': '2020-04-01',
     'still_on_page': True,
     'uri': 'http://jaympatel.com/about'},
    {'domain': 'jaympatel.com',
     'extracted_on': '2018-12-28',
     'last_seen_on': '2019-02-08',
     'still_on_page': False,
     'uri': 'http://jaympatel.com'}],
   'twitter': None,
   'type': 'personal',
   'value': 'j**@jaympatel.com'}],
 'organization': None,
 'pattern': '{first}',
 'state': None,
 'webmail': False},
'meta': {'limit': 5,
 'offset': 0,
```

```
 'params': {'company': None,
  'department': None,
  'domain': 'jaympatel.com',
  'seniority': None,
  'type': None},
 'results': 1}}
```

Let us ignore metadata such as person name, person organization, LinkedIn URL, and so on and only focus on the fact that Hunter has figured out an effective way to perform bulk search for email addresses and their source URLs. Listing 4-2 cleans up the JSON and only prints the email address and source URLs.

Listing 4-2. Cleaning up email addresses

```
temp_dict = json.loads(html_response)
print("email: ", temp_dict["data"]["emails"][0]["value"])
print("\nsource urls:\n")
for i in range(len(temp_dict["data"]['emails'][0]["sources"])):
    print(temp_dict["data"]['emails'][0]["sources"][i]["uri"])
```

```
# Output

email:   j**@jaympatel.com

source urls:

http://leanpub.com/getting-structured-data-from-internet-web-scraping-and-
rest-apis/email_author/new
http://leanpub.com/u/jaympatel
http://jaympatel.com/cv
http://jaympatel.com/consulting-services
http://leanpub.com/getting-structured-data-from-internet-web-scraping-and-
rest-apis
http://jaympatel.com/about
http://jaympatel.com
```

A great way to replicate this functionality is by using regex to extract email addresses. Let us try to do it while crawling through warning letters from the US FDA website. We generated the warning letters CSV file from a table in Chapter 2, and it contained relative link URLs for individual warning letters.

We are loading the email_list into a pandas dataframe to handle the duplicates generated from a special case where a particular email address is present multiple times on a given page as shown in Listing 4-3. In the next chapter, you will learn about how to do the same by simply loading it into a SQL database.

Listing 4-3. Loading email addresses into a dataframe

```
import numpy as np
import pandas as pd
import tld

#df = pd.read_csv("us_fda_url.csv")
df = pd.read_csv("warning_letters_table.csv")
df.head()
fetch_list = []
for i in range(len(df["path"])):
    temp_url = "https://www.fda.gov" + df["path"].iloc[i]
    fetch_list.append(temp_url)

import re
import requests
import tld
import time
def extract_emails(html_res,url, email_list):

    reg = re.compile("([a-zA-Z0-9_.+-]+@[a-zA-Z0-9-]+\.[a-zA-Z0-9-.]+)")
    email_match = reg.findall(html_res)
    for email in email_match:
        potential_tld = "http://"+email.split('@')[1]

        try:
            res = tld.get_tld(potential_tld)
```

```python
        except:
            continue

        temp_dict = {}
        temp_dict["email"] = email
        temp_dict["url"] = url
        email_list.append(temp_dict)
    return email_list

def fetch_pages(email_list, url_list):
    total_urls = len(url_list)
    i = 0
    for url in url_list:
        i = i +1
        time.sleep(1)
        r = requests.get(url)
        if r.status_code == 200:

            html_response = r.text
            email_list = extract_emails(html_response,url, email_list)
            print("fetched " + str(i) + " out of total " + str(total_urls)
            + " pages")
        else:
            continue
    return email_list

email_list = []
email_list = fetch_pages(email_list, fetch_list[:30])
df_emails = pd.DataFrame(email_list)
df_emails.email = df_emails.email.str.lower()
df_emails = df_emails.drop_duplicates(subset = ["email","url"])
df_emails.head(5)
```

Output:

	email	url
0	lynn.bonner@fda.hhs.gov	https://www.fda.gov/inspections-compliance-enf...
2	feb@fda.hhs.gov	https://www.fda.gov/inspections-compliance-enf...
4	alan@thepipeshop.co.uk	https://www.fda.gov/inspections-compliance-enf...
5	ctpcompliance@fda.hhs.gov	https://www.fda.gov/inspections-compliance-enf...
7	abuse@webfusion.com	https://www.fda.gov/inspections-compliance-enf...

We can use this information to create a CSV file which basically contains additional metadata which can be used to eventually build the same output as Hunter.io JSON.

Re2 regex engine

The preceding approach works well, but the only sticking point is the relative performance of Python's regex itself.

Regex are usually very fast, and this leads to their widespread use in validating fields in forms and elsewhere. You may be forgiven if you thought that regex performance across programming languages will be relatively efficient considering it has been in use for over a couple of decades. It's far from being the case, especially since most programming languages expanded basic regex to include features such as backtracking support which allows for specifying that an earlier matched group string must be present at the current string location. This can lead to exponential time complexity especially in cases where regex fails to find a match.

It is also known as catastrophic backtracking where the regex fails to match with the input string by going through an explosive number of iterations. These issues lead to exponential runtimes, and malicious exploits based on these are known as the regular expression denial of service (ReDoS). In Listing 4-4, we have set up a regex which searches for any references of x plus any number of text as a group. The string on which this regex operates contains no y so the regex is destined to fail in all cases. Instead of figuring out this failure and stopping in a timely fashion, this badly written regex will try to match as many x as it can before trying to match all of them and look for y.

Listing 4-4. Exponential time complexity for regex

```
import time
import re
for n in range(20,30):
    time_i = time.time()
    s = 'x'*n
    pat = re.compile('(x+)+y')
    re_mat = pat.match(s)
    time_t = time.time()
    print("total time taken: ", time_t-time_i,s)
Output:
total time taken:  0.048684120178222656 xxxxxxxxxxxxxxxxxxxx
total time taken:  0.10327529907226562 xxxxxxxxxxxxxxxxxxxxx
total time taken:  0.20654749870300293 xxxxxxxxxxxxxxxxxxxxxx
total time taken:  0.38903379440307617 xxxxxxxxxxxxxxxxxxxxxxx
total time taken:  0.7499098777770996 xxxxxxxxxxxxxxxxxxxxxxxx
total time taken:  1.3975293636322021 xxxxxxxxxxxxxxxxxxxxxxxxx
total time taken:  2.844536304473877 xxxxxxxxxxxxxxxxxxxxxxxxxx
total time taken:  5.762147426605225 xxxxxxxxxxxxxxxxxxxxxxxxxxx
total time taken:  11.175331115722656 xxxxxxxxxxxxxxxxxxxxxxxxxxxx
total time taken:  22.800831079483032 xxxxxxxxxxxxxxxxxxxxxxxxxxxxx
```

Due to such pitfalls, we need to be super careful in designing our own regex and to properly test it before we put anything to production. Another failsafe way is to switch out the default regex implementation engine in a given programming language to a more restrictive regex engine with no support for backreferences so that we never run into exponential runtimes

Efficient regex engines such as from Intel's hyperscan library are powering GitHub to perform token scanning at commit time to check for any AWS API keys in public repositories and to notify the user and the cloud provider if an API key is found (https://github.blog/2018-10-17-behind-the-scenes-of-github-token-scanning/).

There are many benchmarks available online, but I took the first one I found (http://lh3lh3.users.sourceforge.net/reb.shtml), and it mentions that Python's built-in regex engine takes about 16 ms compared to 3.7 ms for JavaScript and ~7 ms for C++'s Boost libraries for running the email regex. However, we also notice that finite state machine–based regex engine libraries such as Google's Re2 library run the same regex in only 0.58 ms.

We can test the performance of Re2 ourselves using a Python binding–based high-level library called cffi_re2. You will need to install the re2 library before installing the Python library; and doing it might take some fiddling based on the operating system of your local system.

In the case of Ubuntu, all you need to do is install the re2 library, download the build tools, and finally install the cffi-re2 bindings using the pip command.

```
sudo apt-get install libre2-dev
sudo apt-get install build-essential
pip install cffi_re2
```

If your OS is something other than Ubuntu, you can try to replicate the preceding steps. I recommend simply spinning up an EC2 micro instance with Ubuntu. You can go to launch instance on the EC2 dashboard, text search for "anaconda," in the filters on the left pane, and select Ubuntu; you should see anaconda3-5.1.0-on-ubuntu-16.04-lts – ami-47d5e222 in the community APIs section, select that, and pick t2.micro if you want to replicate exactly the same conditions as follows.

We want to compare the performance of regex engines themselves so it's better if we fetch all the HTML pages first and load them in a CSV file before running regex on them as shown in Listing 4-5.

Listing 4-5. Loading HTML pages into a CSV file

```
# get html of each page
import time
import requests
import numpy as np
import pandas as pd

html_list = []

def fetch_pages(url_list, html_list):
    total_urls = len(url_list)
    i = 0
    for url in url_list:
        i = i +1
```

```
        time.sleep(1)
        r = requests.get(url)
        if r.status_code == 200:

            html_response = r.text
            html_list.append(html_response)
            print("fetched " + str(i) + " out of total " + str(total_urls)
            + " pages")
        else:
            continue
    return html_list

html_list = fetch_pages(fetch_list[:30], html_list)

df_html = pd.DataFrame({'url':fetch_list[:30], 'html': html_list})
df_html.to_csv("us_fda_raw_html.csv")
df_html.head(1)
```

Output

	html	url
0	<!DOCTYPE html>\n<html lang="en" dir="ltr" pr...	https://www.fda.gov/inspections-compliance-enf...

So we now have a file with URLs and HTML source loaded up as a CSV file; we can iterate through this in the script shown in Listing 4-6 and save the total time for up to 640 iterations.

Listing 4-6. Comparing Python's regex engine with re2

```
# save it as a .py file and run it on a machine with re2 and python library
(cffi-re2) installed on it
import re
import tld
import time

import pandas as pd
import numpy as np
import cffi_re2
```

```
import time

def extract_emails(html_list, url_list, reg):

    email_list = []
    for i, html_res in enumerate(html_list):
        email_match = reg.findall(html_res)
        for email in email_match:
            potential_tld = "http://"+email.split('@')[1]

            try:
                res = tld.get_tld(potential_tld)
            except:
                continue

            temp_dict = {}
            temp_dict["email"] = email
            temp_dict["url"] = url_list.iloc[i]
            email_list.append(temp_dict)

    return email_list

def profile_email_regex(reg, iterations, df_html):

    python_engine_list = []
    for iteration in iterations:
        start_time = time.time()
        for i in range(iteration):
            email_list = extract_emails(df_html["html"], df_html["url"], reg)

        end_time = time.time()
        total_time = end_time-start_time
        python_engine_list.append(total_time)
        print("total time (in seconds) for " + str(iteration) + " is ",
        end_time-start_time)
    return email_list, python_engine_list
```

```
if __name__ == "__main__":  # confirms that the code is under main function

    df_html = pd.read_csv("/home/ubuntu/server_files/us_fda_raw_html.csv")

    iteration_list = [10,20,40,80,160,320,640]
    reg = re.compile("([a-zA-Z0-9_.+-]+@[a-zA-Z0-9-]+\.[a-zA-Z0-9-.]+)")
    print("profiling Python 3 regex engine\n")
    email_list_py, python_engine_list = profile_email_regex(reg, iteration_
    list, df_html)

    print("profiling re2 regex engine\n")
    reg = cffi_re2.compile("([a-zA-Z0-9_.+-]+@[a-zA-Z0-9-]+\.
    [a-zA-Z0-9-.]+)")
    email_list_re2, re2_engine_list = profile_email_regex(reg, iteration_
    list, df_html)

    df_emails_re2 = pd.DataFrame(email_list_re2)
    df_emails_re2.to_csv("/home/ubuntu/server_files/emails_re2.csv")

    df_emails_py = pd.DataFrame(email_list_py)
    df_emails_py.to_csv("/home/ubuntu/server_files/emails_py.csv")

    df_profile = pd.DataFrame({"iteration_no":iteration_list, "python_
    engine_time": python_engine_list, "re2_engine_time": re2_engine_list})
    df_profile.to_csv("/home/ubuntu/server_files/profile.csv")
```

Note that each iteration is going through 30 HTML files, so in fact we are measuring how long it takes to go through about 19,000 HTML pages. Another way we typically measure performance is looking at the number of GB processed. At 640 iterations, we have roughly processed 900 MB which is pretty nontrivial especially for a t2.micro EC2 instance which is one of the cheapest EC2 instances.

We can look at the results in Listing 4-7 that the re2 regex engine is about 7–8X faster with only about 8 seconds for going 640 iterations.

Listing 4-7. Printing the comparison table for Python and re2 regex engines

```
df = pd.read_csv("profile.csv", index_col = 'Unnamed: 0')
df.head(10)
Output:
```

	iteration_no	python_engine_time	re2_engine_time
0	10	1.053999	0.135478
1	20	1.988600	0.267462
2	40	4.009065	0.538043
3	80	8.073758	1.066098
4	160	16.062259	2.134781
5	320	31.771234	4.313386
6	640	63.681975	8.589288

We can make a bar plot for better visually comparing the Python regex engine against re2 as shown in Figure 4-1.

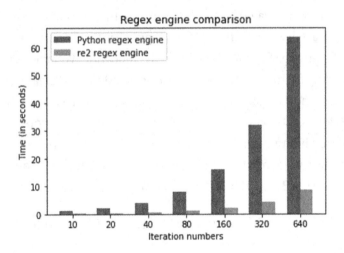

Figure 4-1. *Regex engine comparison*

Named entity recognition (NER)

We have seen that regex allows you to extract specific pieces of information based on matching a pattern of characters or words. This is effective for information with a built-in syntactic structure such as an email address, phone number, and so on or even things such as genetic structures or chemical structures.

However, we need to use statistical and machine learning methods to extract out specific tokens (words or phrases) from unstructured text belonging to specific categories such as names of persons, companies, geographical locations, and so on.

These are referred to as named entity recognition (NER), and it consists of two separate tasks, the first being text segmentation (similar to chunking) where a "name" is extracted and the second classifying it within predefined categories.

Pretrained language models in popular libraries such as Stanford CoreNLP, Spark NLP, and SpaCy all package a NER model which recognizes entities such as person, organization, geographical locations, and so on.

One of the advantages of regex is that it usually works pretty well on raw HTML responses so we don't spend computational resources in trying to parse text out of it. However, NER models generally cannot work on that, and you'll only get intelligible output if the input text is as clean as possible with no HTML tags, special characters, and so on.

We'll learn about sophisticated boilerplate removal methods in later chapters, but for now, a simpler approach is to parse the HTML response into a Beautifulsoup object and remove scripts and style tags from it as shown in the following. We have removed Unicode characters by encoding them as ASCII; we can perform additional cleaning steps using a regex, but usually this much preprocessing should make the text suitable enough for a NER.

In Listing 4-8, let us use a pretrained SpaCy model called "en_core_web_sm"; it can be easily downloaded by "python -m spacy download en_core_web_sm". There are other pretrained English language models available such as en_core_web_md and en_core_web_lg, but they are 91 MB and 789 MB, respectively.

Listing 4-8. Exploring SpaCy's named entity recognition model

```
import pandas as pd
import numpy as np
import spacy
import json
import time
from bs4 import BeautifulSoup
```

```python
def get_full_text(doc):
    soup = BeautifulSoup(doc, 'html.parser')
    for s in soup(['script', 'style']):
        s.extract()
    return (soup.text.strip()).encode('ascii', 'ignore').decode("utf-8")

df = pd.read_csv("us_fda_raw_html.csv")
df["full_text"] = df["html"].apply(get_full_text)

nlp = spacy.load('en_core_web_sm')
ner_list = []
for i,document in enumerate(df["full_text"]):
    start_time = time.time()
    doc = nlp(document)

    person_list = []
    org_list = []
    for ent in doc.ents:

        if ent.label_ == 'PERSON':
            person_list.append(str(ent).lower())
        if ent.label_ == 'ORG':
            org_list.append(str(ent).lower())
    end_time = time.time()
    temp_dict = {}
    temp_dict["url"] = df["url"].iloc[i]
    temp_dict["persons"] = json.dumps(person_list)
    temp_dict["orgs"] = json.dumps(org_list)
    ner_list.append(temp_dict)
    print("Total time (in sec) for iteration number "+ str(i)+ " was " +
    str(end_time-start_time))
```

#output
```
Total time (in sec) for iteration number 0 was 0.6266412734985352
Total time (in sec) for iteration number 1 was 0.2737276554107666
Total time (in sec) for iteration number 2 was 0.31283044815063477

....
(Output truncated)
```

151

We can see that it's taking 0.2–0.6 seconds to go through each document and extract the names. This is not atypical at all; NERs are orders of magnitude more computationally intensive than a well-written regex, even more so if you factor in time for cleaning text which we didn't do for regex.

In my experience, it usually makes sense to use a regex first in a data processing pipeline and only send those documents over to NER which meets some regex condition.

The output from a pretrained NER is pretty noisy, as seen in Listing 4-9, especially if the inference document is not similar to the text corpus on which NER was trained. There are two main ways of dealing with this noise; the easy method is simply to write a postprocessing script which removes junk using a rule-based method. A more robust approach is to train a new NER model using training data which is very similar to inference documents. We will go through both these approaches in the following.

Listing 4-9. Output from SpaCy's NER

```
df_ner = pd.DataFrame(ner_list)
df_ner.persons.iloc[0]
#Output
'["skip", " \\n\\n\\n\\n\\n\\n", "\\n\\n\\nsubmit", "tobacco retailer",
"\\n\\n\\n\\n\\nshare", "jason s. christoffersen", "christoffersen",
"lacf", "lacf", "gourmet gravy", "lot", "cook", "catch 5.5", "retort",
"giblets", "giblets", "giblets", "atids", "b)(4", "lot", "lot", "lot",
"lot", "lot", "dog lamb", "rice", "lot", "lot", "rice entre", "lot", "lot",
"lot", "lot", "lot", "giblets", "lot", "lynn s. bonner", "compliance
officer", "compliance officer bonner", "anne e. johnson", "district\\n\\
n\\n", "w. patrick mcginnis", "john bear", "lydia johnson", "\\n\\n\\n\\
n\\n\\nfda"]'
```

Our end goal is to extract full names corresponding to extracted email addresses from our regex; we can write a script which extracts only those names with email addresses from NER as shown in Listing 4-10.

Listing 4-10. Cleaning the NER output

```
import spacy
import re
import json
import numpy as np
import pandas as pd

df_emails = pd.read_csv("emails_re2_duplicates_removed.csv")

def ner(entities_df, email_df):
    email_person_list = []
    for i,url in enumerate(email_df["url"]):

        email = df_emails.email.iloc[i]

        local_email_name = email.split('@')[0]
        #print(local_email_name)
        email_name_list = re.split('[.,_,-]',local_email_name)

        for name in email_name_list:
            person_list = json.loads(entities_df[entities_df["url"] ==
            url].persons.iloc[0])
            for person in person_list:

                if name in person:
                    temp_dict = {}
                    temp_dict["email"] = email
                    temp_dict["person"] = person
                    temp_dict['url'] = url
                    email_person_list.append(temp_dict)
                    break

    return email_person_list
email_person_list = ner(df_ner, df_emails)
#print(email_person_list)
pd.DataFrame(email_person_list).drop_duplicates().reset_index(drop=True)
```

#Output

email	person	url
0	lynn.bonner@fda.hhs.gov	lynn s. bonner	`https://www.fda.gov/` `inspections-compliance-enf...`
1	alan@thepipeshop.co.uk	alan myerthall	`https://www.fda.gov/` `inspections-compliance-enf...`
2	lillian.aveta@fda.hhs.gov	lillian c. aveta	`https://www.fda.gov/` `inspections-compliance-enf...`
3	matthew.dionne@fda.hhs.gov	matthew r. dionne	`https://www.fda.gov/` `inspections-compliance-enf...`
4	yvette.johnson@fda.hhs.gov	yvette johnson	`https://www.fda.gov/` `inspections-compliance-enf...`
5	robin.rivers@fda.hhs.gov	robin m. rivers	`https://www.fda.gov/` `inspections-compliance-enf...`
6	araceli.rey@fda.hhs.gov	araceli rey	`https://www.fda.gov/` `inspections-compliance-enf...`
7	robin.rivers@fda.hhs.gov	robin m. rivers	`https://www.fda.gov/` `inspections-compliance-enf...`

Training SpaCy NER

SpaCy allows you to train their convolutional neural network (CNN)–based NER model on preloaded entities (person, organizations, etc.) or new entities by using an annotated training dataset using the format shown in Listing 4-11. Training datasets are used in supervised learning to let the algorithm learn from the data by fitting appropriate parameters which enable it to make predictions on unseen data.

It's generally a good idea to at least start with 200–400 annotated sentences as a training set; here we will only use a handful to demonstrate a training approach. Training dataset in SpaCy's simple training style has to be annotated with start and end character indexes of named entities of interest as shown in the following. Doing this manually by hand is almost impossible on large-sized documents. Hence, there are lots of annotation tools such as Brat (`https://brat.nlplab.org/`), WebAnno (`https://webanno.github.io/webanno/`), and Prodigy (`https://prodi.gy/`) on the market today.

Listing 4-11. SpaCy simple training style

```
training_data = [
    (
        'Uber was forced to pay $20m to settle allegations that the company
        duped people into driving with false promises about earnings.',
        {'entities': [(0, 4, 'COM')]}
    )
]
```

All of these tools have their own advantages and disadvantages. Prodigy is the most feature rich and pretty well integrated with SpaCy since it's developed by the same company; however, it's not free, and commercial licenses cost about $490 per seat with a five-seat minimum.

At Specrom Analytics, we use our own in-house developed annotator for NER, but I will show you how we can use a simple word file as an annotator here for SpaCy which should be pretty useful if you don't plan to annotate thousands of documents.

We have loaded the docx file with sentences to be used as a training set; and for annotations, we have simply highlighted the named entities of interest (name of companies) with yellow color. Let us use the docx package (pip install python-docx) in Listing 4-12 to get a list of all the named entities from the training set.

Listing 4-12. Using the docx Python package

```
from docx import Document
```

```
from docx.shared import Inches
from docx.enum.text import WD_COLOR_INDEX
```

```
entities_yellow_set = set()
```

```
document = Document('training_sample.docx')
for paragraph in document.paragraphs:
    for run in paragraph.runs:

        if run.font.highlight_color is not None:

            #print(run.text, run.font.highlight_color)
            pass
        if run.font.highlight_color == WD_COLOR_INDEX.YELLOW:
            entities_yellow_set.add(run.text)

entities_list = list(entities_yellow_set)
print(entities_list)

# Output

['Waymo', 'Uber's', 'Uber', 'Alphabet', 'Berkshire Hathaway', 'Google's',
'General Electric']
```

Now, all we need is to write some helper code in Listing 4-13 to find these entities and their start and stop indexes and format the output into SpaCy simple training data format.

Listing 4-13. Converting to SpaCy's simple training style

```
def get_start_stop_index(string_long, substring):
    len2 = len(substring)
    start_index = string_long.find(substring)
    end_index = start_index+len2
    return start_index, end_index

return_list = []
#for k in range(len(country_yellow_set)):
for paragraph in document.paragraphs:
    individual_return_tuple = []
    #print(paragraph.text)
    individual_return_tuple.append(paragraph.text.strip())
    entity_return_list = []
    return_entity_dict = {}
```

```
    for entity in entities_list:

        #entity_return_list = []

        if entity in paragraph.text:
            #print(entity)
            #return_tuple = []
            #print(paragraph.text)
            #return_tuple.append(paragraph.text)
            start_in, stop_in = get_start_stop_index(paragraph.text, entity)

            entity_return = [start_in, stop_in, "COM"]
            entity_return = tuple(entity_return)
            entity_return_list.append(entity_return)
            #print(entity_return_list)
    if len(entity_return_list) != 0:
        return_entity_dict["entities"] = entity_return_list
        #print(return_entity_dict)
        individual_return_tuple.append(return_entity_dict)
        individual_return_tuple = tuple(individual_return_tuple)
        return_list.append(individual_return_tuple)
        #print(get_start_stop_index(paragraph.text, country))
    #print("*"*20)
return_list

# Output

[('Uber was forced to pay $20m to settle allegations that the company duped
people into driving with false promises about earnings.',
  {'entities': [(0, 4, 'COM')]}),
 ('The Federal Trade Commission claimed that most Uber drivers earned far
  less than the rates Uber published online in 18 major cities in the US.',
  {'entities': [(47, 51, 'COM')]}),
 ('Former Uber engineer Susan Fowler went public with allegations of sexual
  harassment and discrimination, prompting the company to hire former US
  attorney general Eric Holder to investigate her claims.',
  {'entities': [(7, 11, 'COM')]}),
```

```
('Waymo, the self-driving car company owned by Google's parent corporation
 Alphabet, filed a lawsuit against Uber, accusing the startup of
 "calculated theft" of its technology.',
 {'entities': [(0, 5, 'COM'),
   (107, 111, 'COM'),
   (73, 81, 'COM'),
   (45, 53, 'COM')]}),
('The suit, which could be a fatal setback for Uber's autonomous vehicle
 ambitions, alleged that a former Waymo employee, Anthony Levandowski,
 stole trade secrets for Uber.',
 {'entities': [(104, 109, 'COM'), (45, 51, 'COM'), (45, 49, 'COM')]}),
('Uber later fired the engineer.', {'entities': [(0, 4, 'COM')]}),
('It is easiest to think of the firm as a holding company, lying somewhere
 between Warren Buffet's private equity firm Berkshire Hathaway and the
 massive conglomerate that is General Electric.',
 {'entities': [(117, 135, 'COM'), (173, 189, 'COM')]}),
('Like the former, Alphabet won't have any consumer facing role itself,
 instead existing almost as an anti-brand, designed to give its
 subsidiaries room to develop their own identities.',
 {'entities': [(17, 25, 'COM')]})]
```

Once we have the training data, we can train a new NER model in SpaCy using a blank English language model. We have used the recommended training parameters in the SpaCy documentation such as the number of iterations, minibatch size, and so on in Listing 4-14, but these should all be optimized for your training set.

Listing 4-14. Training SpaCy's NER model

```
import random
from pathlib import Path
import spacy
from spacy.util import minibatch, compounding

# new entity label
label_list = ["COM"]
```

```python
def train_ner(TRAIN_DATA,label_list,output_model_name, model=None,
n_iter=30):
    random.seed(0)
    if model is not None:
        nlp = spacy.load(model)
        print("Loaded model '%s'" % model)
    else:
        nlp = spacy.blank("en")
        print("Created blank 'en' model")

    if "ner" not in nlp.pipe_names:
        ner = nlp.create_pipe("ner")
        nlp.add_pipe(ner)
    else:
        ner = nlp.get_pipe("ner")
    for LABEL in label_list:
        ner.add_label(LABEL)
    nlp.vocab.vectors.name = 'spacy_pretrained_vectors'
    if model is None:

        optimizer = nlp.begin_training()
    else:

        optimizer = nlp.resume_training()
    move_names = list(ner.move_names)

    pipe_exceptions = ["ner", "trf_wordpiecer", "trf_tok2vec"]
    other_pipes = [pipe for pipe in nlp.pipe_names if pipe not in
    pipe_exceptions]
    with nlp.disable_pipes(*other_pipes):  # only train NER
        sizes = compounding(1.0, 4.0, 1.001)
        for itn in range(n_iter):
            random.shuffle(TRAIN_DATA)
            batches = minibatch(TRAIN_DATA, size=sizes)
            losses = {}
```

```
            for batch in batches:
                texts, annotations = zip(*batch)
                nlp.update(texts, annotations, sgd=optimizer, drop=0.35,
                losses=losses)
            print("Losses", losses)
    nlp.to_disk(output_model_name)
train_ner(TRAIN_DATA = return_list, label_list = label_list, output_model_
name = 'ner_company', model = None,n_iter=30)
# Output
Created blank 'en' model
Losses {'ner': 73.29305328331125}
Losses {'ner': 28.609920230269758}
Losses {'ner': 22.84507976165346}
Losses {'ner': 17.49904894383679}
Losses {'ner': 11.414351243276446}
Losses {'ner': 9.101663951377668}
Losses {'ner': 5.267494453267735}
Losses {'ner': 0.5274157429486094}
Losses {'ner': 5.515589335372427}
Losses {'ner': 1.2058509063100247}
Losses {'ner': 0.034352628598915316}
Losses {'ner': 3.998464111365581}
Losses {'ner': 2.045643676879328}
Losses {'ner': 0.2544095753072009}
Losses {'ner': 3.1612663279231046}
Losses {'ner': 2.0002408098323947}
Losses {'ner': 2.3651013551862197}
Losses {'ner': 3.9945953801627176}
Losses {'ner': 5.44444397987547984}
Losses {'ner': 3.9993801117102756}
Losses {'ner': 1.999933719635059}
Losses {'ner': 0.00016759568611773816}
Losses {'ner': 7.737405588428668e-05}
Losses {'ner': 1.999963760379788}
Losses {'ner': 7.545372597637903e-15}
```

```
Losses {'ner': 4.697406133102387e-09}
Losses {'ner': 0.0011829053454656055}
Losses {'ner': 0.014950245941714796}
Losses {'ner': 1.999995946927806}
Losses {'ner': 3.557686596887918e-07}
```

Let us test the newly trained NER model on a test sentence in Listing 4-15.

Listing 4-15. Testing SpaCy's NER

```
test_text = "SoftBank is often described as the Berkshire Hathaway of tech.
That was once a flattering comparison. But the investing track records for
the Japanese firm run by Masayoshi Son and Berkshire's Warren Buffett have
soured lately."
nlp2 = spacy.load("ner_company")
doc2 = nlp2(test_text)
for ent in doc2.ents:
    if ent.label_ == 'COM':
        print(ent.label_, ent.text)
# Output
COM Berkshire Hathaway
COM Masayoshi Son
```

So it correctly identified Berkshire Hathaway as a company name entity but missed out on Softbank and incorrectly classified Masayoshi Son as a company name. This was to be expected though, since our training set was far too small for handling real-world documents, but I hope this illustrated how easy it is to train a CNN-based NER model in SpaCy.

The real painpoint and a bottleneck is putting together an appropriate NER training dataset since it's more art than science. You typically want the dataset to be large enough to generalize well with your intended application, but it can't be too specific for a problem at hand or you will run into a catastrophic forgetting problem (https://explosion.ai/blog/pseudo-rehearsal-catastrophic-forgetting) where the model may perform well for small use cases but "forgets" about other entities which you had trained in earlier iterations. In a worst-case scenario, we see that sometimes retraining a saved model on new training data just isn't effective, and the only option in such cases is to start over with a new model. You'll also see lots of training recipes and empirical approaches on Kaggle and elsewhere which can be pretty effective if used correctly.

Another way to create a faster training dataset is relying on lookup tables or a terminology list with some rule-based matching using parts of speech (POS) tags either as part of SpaCy's matcher class or separately to annotate named entities. This will quickly generate annotated datasets in cases where rules are easy to define and you have a way to get a terminology list without much effort. For example, in the training NER, instead of manually labeling names of companies, we could've used a lookup table created from a list of all publicly traded companies.

Even this is not entirely effective when we are trying to label entities which are either novel (such as labeling research papers) or not well defined to a layperson. At Specrom Analytics, we employ subject matter experts when annotating training datasets from highly specialized domain areas such as medical, pharmaceutical, legal, and financial areas since the domain area entities are composed of vocabularies which are intelligible only to someone from that field. Examples of domain-specific entities are descriptions of medical problems, names of medical conditions, pharmaceutical or chemical names, and so on.

In a way, this is similar to developing knowledge-based systems (or expert systems) but with much faster turnaround time.

It may seem like an expensive way to label data, but remember that these neural network models operate solely on the quality of training data; if the named entities weren't labeled properly to begin with, there is no way they will give you a good result at inference.

There are excellent pretrained general-purpose NER models available within open source libraries, but due to the high cost of training and testing a domain-specific NER, almost all good domain-specific NER models such as from John Snow Labs (`www.johnsnowlabs.com/spark-nlp-health/`) are available only with an expensive commercial license.

Exploratory data analytics for NLP

In the NER section earlier, SpaCy abstracted away the process of converting words and text documents into numerical vectors through a process known as vectorization.

However, we will have to learn about how to vectorize documents to glean a higher-order meaning about the structure and type of the document.

One of the simplest methods of text vectorization is known as the bag-of-words model where we count the number of unique words or tokens in a document by disregarding grammar, punctuation, and word order.

We will learn about how to preprocess text corpus and group them using topic modeling as well as text clustering algorithms.

Once we have documents labeled into specific groups, then we can treat them as a supervised learning problem and apply text classification algorithms. A classic example of text classification is classifying a given email into spam or not spam based on the text in the email itself. Other examples of text classification include classifying the topic of the news article into financial, entertainment, sports, and so on.

Let us download (`http://mlg.ucd.ie/datasets/bbc.html`) the BBC dataset (Citation: D. Greene and P. Cunningham. "Practical Solutions to the Problem of Diagonal Dominance in Kernel Document Clustering," Proc. ICML 2006.) consisting of about 2200 raw text documents from the BBC News website taken from 2005 to 2006.

The text files are named as 001.txt, 002.txt, and so on; the first line of each file represents the document title. Each file is put into one of the folders labeled as business, entertainment, politics, sports, and tech. Due to the structure of data files, it will take us some data wrangling to get all the text data into a pandas dataframe.

Let's read these text files into a pandas dataframe in Listing 4-16 to perform some preliminary exploratory data analytics (EDA).

Listing 4-16. Creating a dataframe from the BBC News dataset

```
import os

directory = []
file = []
title = []
text = []
label = []
datapath = r'yourfilepath\bbc-fulltext\bbc'
for dirname, dir2 , filenames in os.walk(datapath):

    for filename in filenames:
        directory.append(dirname)
        file.append(filename)
        label.append(dirname.split('\\')[-1])
        #print(filename)
        fullpathfile = os.path.join(dirname,filename)
```

```
    with open(fullpathfile, 'r', encoding="utf8", errors='ignore') as
    infile:
        intext = ''
        firstline = True
        for line in infile:
            if firstline:
                title.append(line.replace('\n',''))
                firstline = False
            else:
                intext = intext + ' ' + line.replace('\n','')
        text.append(intext)
df = pd.DataFrame({'title':title, 'text': text, 'label':label })
df.to_csv("bbc_news_data.csv")
df.head()
Output:
```

	label	text	title
0	business	Quarterly profits at US media giant TimeWarn...	Ad sales boost Time Warner profit
1	business	The dollar has hit its highest level against...	Dollar gains on Greenspan speech
2	business	The owners of embattled Russian oil giant Yu...	Yukos unit buyer faces loan claim
3	business	British Airways has blamed high fuel prices ...	High fuel prices hit BA's profits
4	business	Shares in UK drinks and food firm Allied Dom...	Pernod takeover talk lifts Domecq

Next, we should check if text belonging to different labels is relatively well balanced; if it's not, then we will get a badly trained classification model. Let us plot label counts as the first step in EDA in Listing 4-17 and Figure 4-2. This dataset originated from a published paper; hence, quite predictably, it is already pretty well balanced.

Listing 4-17. Plotting the topic distribution

```
import matplotlib.pyplot as plt
import seaborn as sns

sns.set(style="darkgrid")
f, axes = plt.subplots(1, 1, figsize=(7, 7), sharex=True)
sns.despine(left=True)
```

```
# this creates bar graph for one column called "topic"
col_count_2 = df['label'].value_counts()
sns.set(style="darkgrid")
sns.barplot(col_count_2.index, col_count_2.values, alpha=0.9)
plt.title('Frequency Distribution of Topics', fontsize=13)
plt.ylabel('Number of Occurrences', fontsize=13)
plt.xlabel('Topics', fontsize=13)

plt.xticks(rotation=70, fontsize=13)

plt.setp(axes, yticks=[])
plt.tight_layout()
plt.show()
```

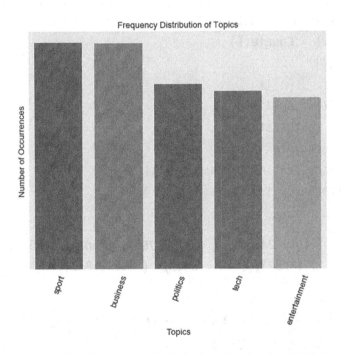

Figure 4-2. *Frequency distribution of labeled topics in the BBC dataset*

Tokenization

The process of converting text into individual words (called tokens) is known as tokenization. This is a necessary first step before we can explore top words in a corpus or perform any other preprocessing steps for text vectorization.

One of the simplest tokenization strategies is simply splitting each word at spaces, and this is referred to as "white space tokenization." There are obvious shortcomings of such an approach, but let us keep that aside and tokenize the corpus using the string split method.

There are sophisticated methods for creating a bag-of-words model under sklearn called countvectorizer, but for now let's stick to using Counter from Python's collections in Listing 4-18 to keep track of word frequencies and use its most_common method to call for top words in the corpus. We will switch to other approaches in forthcoming sections.

Listing 4-18. Querying for top words in text

```
# top words in text

from collections import Counter
termfrequency_text = Counter()
texts = df["text"]
for text in texts:
    text_list = text.split(' ')
    for token in text_list:
        termfrequency_text[token] +=1
print(termfrequency_text.most_common(10))
print(len(termfrequency_text.keys()))
```

Output:
```
[('the', 44432), ('to', 24460), ('of', 19756), ('and', 17867), ('a', 17115),
('in', 16316), ('', 13187), ('is', 8427), ('for', 8424), ('that', 7528)]
64779
```

We can plot a wordcloud in Listing 4-19 using these term frequencies to visualize top words in the corpus. To our dismay, the top words in Figure 4-3 are all stop words such as "are," "and," "the," and so on which aren't really helpful in gleaning any semantic meaning of the text corpus. We also noticed that there are about 59,968 words in the corpus vocabulary; let us apply successive operations in order to reduce the overall size of this vocabulary so that we can have effective vectors representing individual documents.

Listing 4-19. Generating wordclouds

```
from wordcloud import WordCloud, STOPWORDS
wordcloud = WordCloud(background_color="white").generate_from_frequencies
(frequencies=termfrequency_text)
# Generate plot
plt.imshow(wordcloud)
plt.axis("off")
plt.show()
```

Figure 4-3. *Wordcloud for top words in the BBC dataset*

Advanced tokenization, stemming, and lemmatization

Many languages including English use word contractions in spoken and written speech, for example, two words "can" and "not" are frequently contracted to "can't," do + not into "don't," and I + have into "I've." Our tokenization algorithm should be able to take care of this; otherwise, we will end up with a nondictionary word "cant" once we remove the punctuation.

NLTK's treebank tokenizer splits word contractions into two tokens as in Listing 4-20.

Listing 4-20. Treebank tokenizer

```
from nltk.tokenize import TreebankWordTokenizer
sample_text = "can't don't won't I've running run ran"
def tokenizer_tree(sample_text):
    tokenizer = TreebankWordTokenizer()
    tokenized_list= tokenizer.tokenize(sample_text)
    return tokenized_list
```

```
tokenizer_tree(sample_text)
# Output:
['ca', "n't", 'do', "n't", 'wo', "n't", 'I', "'ve", 'running', 'run', 'ran']
```

SpaCy's tokenization also produces a similar result in Listing 4-21.

Listing 4-21. SpaCy tokenization example

```
sample_text =  "can't don't won't I've running run ran"
from spacy.lang.en import English
nlp = English()
tokenizer = nlp.Defaults.create_tokenizer(nlp)
tokens = tokenizer(sample_text)
token_list = []
for token in tokens:
    token_list.append(token.text)
token_list
# output
['ca', "n't", 'do', "n't", 'wo', "n't", 'I', "'ve", 'running', 'run', 'ran']
```

Once we separate out the word contractions, we still need to convert different word inflections such as run, ran, and running into the root word. One of the ways to do it in a computationally cheap way is known as stemming where the algorithm basically chops off the last few characters of some words based on a heuristic. This may result in some nondictionary words such as double being converted to doubl, but generally speaking, this does reduce the size of the vocabulary very well as demonstrated in Listing 4-22.

Listing 4-22. Porter and snowball stemmer

```
from nltk.stem.porter import PorterStemmer
def stemmer_porter(text_list):
    porter = PorterStemmer()
    return_list = []
    for i in range(len(text_list)):
        return_list.append(porter.stem(text_list[i]))
    return(return_list)
```

```
# Another popular stemmer
from nltk.stem.snowball import SnowballStemmer

def stemmer_snowball(text_list):
    snowball = SnowballStemmer(language='english')
    return_list = []
    for i in range(len(text_list)):
        return_list.append(snowball.stem(text_list[i]))
    return(return_list)
print(stemmer_porter(tokenizer_tree(sample_text)))
# Output
['ca', "n't", 'do', "n't", 'wo', "n't", 'I', "'ve", 'run', 'run', 'ran',
'doubl']
```

A much better set of algorithms which handles word inflections is known as
lemmatization; this gives you dictionary words as shown in Listing 4-23 instead of
stemming. One serious drawback is the 2–3X more time requirements over stemming
methods which may make it impractical for many use cases related to web crawl data.

Listing 4-23. SpaCy lemmatization

```
sample_text =  "can't don't won't running run ran double"
from spacy.lang.en import English
def tokenizer_lemma(corpus_text):
    nlp = English()
    tokenizer = nlp.Defaults.create_tokenizer(nlp)
    tokens = tokenizer(corpus_text)
    lemma_list = []
    for token in tokens:
        lemma_list.append(str(token.lemma_).lower())
    return lemma_list
print(tokenizer_lemma(sample_text))
# Output
['can', 'not', 'do', 'not', 'will', 'not', 'run', 'run', 'run', 'double']
```

I personally use stemming wherever the output is not going to be visible to the
end user such as preprocessing text when searching through a database, computing
document clusters or document similarity, and so on.

However, if I am going to plot a wordcloud of the most common text, then it's preferable to do lemmatizations instead so that we can guarantee that the normalized words are actual dictionary words.

I think it's a good time to introduce sklearn's countvectorizer module (Listing 4-24) so that we can leverage its powerful interface and ability to bring in our own custom tokenizer and stemmer or lemmatizer functions.

Listing 4-24. Sklearn's countvectorizer

```
def snowball_treebank(sample_string):
    return stemmer_snowball(tokenizer_tree(sample_string))

from sklearn.feature_extraction.text import CountVectorizer

cv = CountVectorizer(tokenizer=snowball_treebank)
cv_train = cv.fit_transform(df["text"])
df_dtm = pd.DataFrame(cv_train.toarray(), columns=cv.get_feature_names())
print(df_dtm.head())

top_10_count = pd.DataFrame(np.asarray(cv_train.sum(axis=0)),columns=cv.
get_feature_names()).transpose().rename(columns = str).sort_values(by =
'0',axis = 0, ascending = False).head(10)
print(top_10_count)

#Output
```

	!	#	$	%	&	'	"	'.	'd	'm	...	£950,000	£952m	£960m	£96bn	£97m	£98	£980m	£98m	£99	£9m
0	0	0	9	8	0	1	1	0	0	0	...	0	0	0	0	0	0	0	0	0	0
1	0	0	2	0	0	0	0	3	0	0	...	0	0	0	0	0	0	0	0	0	0
2	0	0	4	0	0	1	2	0	0	0	...	0	0	0	0	0	0	0	0	0	0
3	0	0	1	8	0	0	6	0	0	0	...	0	0	0	0	0	0	0	0	0	0
4	0	0	2	3	0	1	0	0	0	0	...	0	0	0	0	0	0	0	0	0	0

5 rows × 32641 columns

	0
the	52542
,	35436
to	24632
of	19886
and	18535
a	18255
in	17409
``	11160
it	9890
''	9282

Each row in the preceding dataframe corresponds to one document; the number in each cell represents the term frequency or number of times the word in the column heading appeared in a particular document. This format of representing text corpus is known as the document term matrix.

We have created a memory-intensive dense array from a sparse matrix by using the .toarray() method, and hence we should only do this when we are certain that we will be able to comfortably fit this dataframe in our memory. An easy way to check that is by calling len(cv.get_feature_names()); if the number of features is a few thousand, then we'll probably be OK in converting it to a pandas dataframe, but for features or a number of rows in hundreds of thousands, it's best to avoid this step.

We'll apply simple dataframe–based manipulations to generate top ten words in the corpus akin to what we saw with the Counter method.

As you can see, all the top words are dominated by either punctuation marks or by common words which don't contribute much to the semantic meaning of the text documents.

Punctuation removal

Let's use a regex-based expression in Listing 4-25 to get rid of punctuation marks from our text corpus. Countvectorizer allows us to pass a preprocessor function which can take care of this particular problem. We will continue to use the same word tokenizer as before.

Listing 4-25. Punctuation removal

```
def preprocessor_final(text):
    if isinstance((text), (str)):
        text = re.sub('<[^>]*>', ' ', text)
        text = re.sub('[\W]+', ' ', text.lower())
        return text
    if isinstance((text), (list)):
        return_list = []
        for i in range(len(text)):
            temp_text = re.sub('<[^>]*>', '', text[i])
            temp_text = re.sub('[\W]+', '', temp_text.lower())
            return_list.append(temp_text)
        return(return_list)

from sklearn.feature_extraction.text import CountVectorizer

cv = CountVectorizer(lowercase=True, preprocessor=preprocessor_final,
tokenizer=snowball_treebank)
cv_train = cv.fit_transform(df["text"])
df_dtm = pd.DataFrame(cv_train.toarray(), columns=cv.get_feature_names())
print(df_dtm.head())
top_10_count = pd.DataFrame(np.asarray(cv_train.sum(axis=0)),columns=cv.
get_feature_names()).transpose().rename(columns = str).sort_values(by =
'0',axis = 0, ascending = False).head(10)
print(top_10_count)
# Output
```

	0	00	000	0001	000bn	000m	000s	000th	001	001and	...	zoom	zooropa	zornotza	zorro	zubair	zuluaga	zurich	zuton	zvona	zvyagireva	ntsev
0	0	0	1	0	0	0	0	0	0	0	...	0	0	0	0	0	0	0	0	0	0	0
1	0	0	0	0	0	0	0	0	0	0	...	0	0	0	0	0	0	0	0	0	0	0
2	0	0	0	0	0	0	0	0	0	0	...	0	0	0	0	0	0	0	0	0	0	0
3	0	0	1	0	0	0	0	0	0	0	...	0	0	0	0	0	0	0	0	0	0	0
4	0	0	0	0	0	0	0	0	0	0	...	0	0	0	0	0	0	0	0	0	0	0

173

5 rows × 20504 columns

	0
the	52574
to	24767
of	19930
and	18574
a	18297
in	17558
it	10171
s	8954
for	8732
is	8534

We have reduced the total tokens to about 20,000, and now all the top ten words are represented by stop words.

Ngrams

So far, we have seen that our bag-of-words model is based on creating tokens out of individual words; this approach is also known as unigram-based tokenization. It's been found empirically we can make a more accurate model by vectorizing text documents using an ngram model.

For example, if we have a sentence "Ask Jeeves has become the third leading online search firm this week to thank a revival in internet advertising for improving fortunes," then the extracted bigrams from the model will be "Ask Jeeves," "Jeeves has," "has become," ... "improving fortunes."

If we would have extracted three words, then the ngram would have been called a trigram and so on. It's typical not to go for higher ngrams since it leads to an exponential increase in the number of tokens.

It's pretty common to combine a unigram and bigram for most use cases. Countvectorizer allows us to do so by specifying the ngrams_range parameter as shown in Listing 4-26.

Listing 4-26. ngrams

```
from sklearn.feature_extraction.text import CountVectorizer

cv = CountVectorizer(lowercase=True, preprocessor=preprocessor_final,
tokenizer=snowball_treebank, ngram_range=(1, 2))
cv_train = cv.fit_transform(df["text"])
df_dtm = pd.DataFrame(cv_train.toarray(), columns=cv.get_feature_names())
df_dtm.head()
Output
```

	00	01	02	03	04sec	19sec 2	...	zurich s	zuton at	zuton	zvonareva reva	zvona 6 has	zvonareva russia	zvonareva struggl	zvonareva tsev	zvyagin tsev	zvyagin the
0	0	0	0	0	0	0	...	0	0	0	0	0	0	0	0	0	0
1	0	0	0	0	0	0	...	0	0	0	0	0	0	0	0	0	0
2	0	0	0	0	0	0	...	0	0	0	0	0	0	0	0	0	0
3	0	0	0	0	0	0	...	0	0	0	0	0	0	0	0	0	0
4	0	0	0	0	0	0	...	0	0	0	0	0	0	0	0	0	0

5 rows × 309261 columns

The top ten words are still the same as in the preceding section; however, our vocabulary size has drastically increased to over 300,000 terms. If our corpus was any larger, then we would risk not fitting it in our memory and getting an error.

Stop word removal

Once we include ngrams in our bag-of-words model, the dimensions and the sparsity of data explode to unmanageable levels which make any kind of statistical modeling impossible due to the curse of dimensionality.

We need to exclude a large number of tokens from our corpus which do not contribute much to the underlying meaning of the documents.

There are four main ways to remove stop words.

Method 1: Create an exclusion list

The simplest method is by creating an exclusion list such as shown in Listing 4-27 which is taken from the sklearn countvectorizer's English stop word set. The NLTK library also provides a common stop word list for the English language.

Listing 4-27. Stop word list

```
stop_words_string = "'less,hers,meanwhile,then,fire,been,couldnt,hundred,
forty,nine,every,over,these,where,all,cannot,due,interest,this,by,yet,
formerly,or,will,fifty,hereupon,again,behind,nor,sincere,thereafter,front,
and,to,whereupon,eight,into,me,somehow,which,must,thick,with,anywhere,co,
mill,once,almost,how,should,first,off,un,since,i,same,an,throughout,however,
one,between,someone,whereafter,during,became,six,ltd,something,often,latter,
find,their,her,whereas,thereby,full,has,still,done,former,our,up,ever,my,
detail,see,third,herself,us,very,myself,describe,there,ourselves,thru,
thence,much,somewhere,moreover,your,perhaps,back,ten,whereby,twelve,have,
via,before,as,mostly,yourselves,name,toward,would,nowhere,enough,sixty,
them,put,yours,therein,if,be,alone,along,anything,do,fill,now,re,made,few,
whose,it,his,seems,is,more,upon,any,amoungst,last,give,otherwise,are,being,
herein,yourself,others,through,namely,becoming,several,also,cry,everything,
```

of,together,towards,five,no,because,con,show,anyway,ie,can,therefore,three,
a,indeed,afterwards,found,hereby,move,itself,amount,please,seemed,out,she,
than,such,amongst,beyond,but,hence,become,ours,so,least,thus,while,every
where,here,bill,anyhow,whether,had,in,eleven,fifteen,on,whom,you,other,
already,neither,above,part,per,else,another,below,get,elsewhere,he,whatever,
who,even,himself,latterly,hereafter,against,many,always,empty,among,whence,
until,beside,twenty,besides,hasnt,side,some,for,never,system,wherever,
might,not,inc,eg,etc,none,whither,him,its,nevertheless,themselves,two,
around,rather,after,they,de,am,further,whole,everyone,thin,within,go,noone,
nothing,sometimes,that,whoever,whenever,seeming,beforehand,across,may,
thereupon,nobody,from,only,we,why,about,either,wherein,call,the,own,those,
under,what,well,top,were,onto,next,becomes,could,each,serious,take,four,
both,seem,when,without,cant,although,mine,sometime,keep,at,down,though,was,
too,except,anyone,bottom,most"'

Let us exclude these words from our term frequencies in Listing 4-28.

Listing 4-28. Excluding stop words from a hardcoded list

```
# top words in text
```

```
stop_words_list = stop_words_string.split(',')
stemmed_stop_words_list = stemmer_snowball(stop_words_list)
cv = CountVectorizer(stop_words = stemmed_stop_words_list,lowercase=True,
preprocessor=preprocessor_final, tokenizer=snowball_treebank)
cv_train = cv.fit_transform(df["text"])
df_dtm = pd.DataFrame(cv_train.toarray(), columns=cv.get_feature_names())
print(df_dtm.head())
```

```
# Output
```

	0	00	000	0001	000bn	000m	000s	000th	001	001and	...	zoom	zooropa	zornotza	zorro	zubair	zuluaga	zurich	zuton	zvona	zvyagin reva	tsev
0	0	0	0	1	0	0	0	0	0	0	...	0	0	0	0	0	0	0	0	0	0	0
1	1	0	0	0	0	0	0	0	0	0	...	0	0	0	0	0	0	0	0	0	0	0
2	0	0	0	0	0	0	0	0	0	0	...	0	0	0	0	0	0	0	0	0	0	0
3	0	0	0	1	0	0	0	0	0	0	...	0	0	0	0	0	0	0	0	0	0	0
4	0	0	0	0	0	0	0	0	0	0	...	0	0	0	0	0	0	0	0	0	0	0

179

5 rows × 20221 columns

	0
s	8954
said	7254
year	3296
mr	3005
peopl	2043
new	1909
time	1656
game	1634
say	1569
use	1568

We can see that the top ten words are now free of stop words such as the, of, and so on that we had in the "Tokenization" section. One obvious issue with hardcoded stop words–based approach is that we need to preprocess the stop word list with the same stemming algorithm that we used for the text corpus itself; otherwise, we will not filter out all the stop words.

This has been a sole method to remove stop words for quite a long time; however, there is an awareness among researchers and working professionals that such one-size-fits-all method is actually quite harmful in learning about the overall meaning of the text, and there are papers out there (www.aclweb.org/anthology/W18-2502.pdf) which caution against this approach.

Method 2: Using statistical language modeling

The second approach relies on the statistical language-specific model to figure out if a certain word is a stop word or not. SpaCy ships with such a model by default, and we can use it as shown in Listing 4-29.

This method will not be completely effective in removing all stop words from a corpus, and hence you may combine this with a hardcoded stop word list.

However, the idea is that your statistical model will be eventually well trained enough to identify all stop words, and the reliance on a hardcoded list will be kept to a minimum. In practice, we rarely use this type of stop word removal on larger text corpuses due to quite a bit of computational overhead with this method.

Listing 4-29. Using SpaCy's model for identifying stop words

```
from spacy.lang.en import English
start_time = time.time()

def tokenizer_lemma_stop_words(corpus_text):
    nlp = English()
    tokenizer = nlp.Defaults.create_tokenizer(nlp)
    tokens = tokenizer(corpus_text)

    lemma_list = []
    for token in tokens:

        if token.is_stop is False:

            lemma_list.append(str(token.lemma_).lower())
    return lemma_list
```

Method 3: Corpus-specific stop words

The third approach uses something known as corpus-specific stop words. Countvectorizer contains a hyperparameter called max_df which allows you to specify a word frequency threshold above which the words are automatically excluded from the document vocabulary. You can specify a float value between 0 and 1 which will allow you to remove words appearing in proportions of documents greater than the threshold value. If you supply an integer, then that represents an absolute count of documents which contain a certain word. Terms with higher than this count are automatically filtered out.

In a similar vein, there are certain words which are so rare across the corpus that they contribute little to the overall semantic meaning. We can remove these words by specifying a min_df parameter which takes values similar to max_df, but here the terms below this threshold are ignored.

Method 4: Using term frequency–inverse document frequency (tf-idf) vectorization

The last approach modifies our vectorizing method itself; count vectorization assigns equal weightage to both stop words aka common words, which occur in majority of documents in a corpus, and other words sometimes known as content words which impart semantic meaning to a particular document and are in general less commonly found across all documents in a corpus.

A much better way to vectorize text documents is by reducing the weightages of stop words and increasing them for content words so that our vectors appropriately represent the semantics of the underlying documents without requiring us to be completely efficient in removing all stop words from a document. This vectorization method is known as the term frequency–inverse document frequency (tf-idf), and we will use it as our primary vectorization method due to its obvious advantages.

Mathematically speaking, term frequency (TF) is the ratio of the number of times a word or token appears in a document (ni,j) compared to the total number of words or tokens in the same document, and it's expressed as follows:

$$TF = \frac{n_{i,j}}{\sum_k n_{i,j}}$$

Inverse document frequency, idf(w), of a given word w is defined as a log of the total number of documents (N) divided by document frequency df_t, which is the number of documents in the collection containing the word w.

$$idf(w) = \log \frac{N}{df_t}$$

Term frequency–inverse document frequency, tf-idf (w), is simply a product of term frequency and inverse document frequency.

$$tf\text{-}idf(w) = TF \times idf(w)$$

Sklearn contains a tf-idf vectorizer method, and it's almost a drop in replacement for the countvectorizer so we can use it along with other parameters such as max_df and min_df, a hardcoded stop word list, as well as a tokenizer method of our choice as shown in Listing 4-30.

It will also be a good idea to split our dataset into a test and train portion so that we can keep aside a portion of our dataset which can be used for validation and testing once we have developed a trained model.

Listing 4-30. tf-idf vectorization

```
from sklearn.model_selection import train_test_split
train, test = train_test_split(df, test_size=0.2)
print("Train df shape is: ",train.shape)
print("Test df shape is: ",test.shape)

from sklearn.feature_extraction.text import TfidfVectorizer

tfidf_transformer = TfidfVectorizer(stop_words='english',
                                    ngram_range=(1, 2),max_df=0.97, min_df =
                                    0.03, lowercase=True, max_features=2500)

X_train_text = tfidf_transformer.fit_transform(train['text'])
df_dtm = pd.DataFrame(X_train_text.toarray(), columns=tfidf_transformer.
get_feature_names())
df_dtm.head()

# Output

Train df shape is:  (1780, 4)
Test df shape is:  (445, 4)
```

	000 10	100	11 12	13 14 15 16	17 ...	world worth wide	worth	written	wrong year	year	year old	years years ago york	young
0	0.187213	0.000000	0.0 0.000000	0.0 0.0 0.0 0.000000	0.0 0.0 ...	0.000000	0.000000	0.0	0.000000	0.000000	0.064787	0.059774	0.0
1	0.000000	0.000000	0.0 0.000000	0.0 0.0 0.0 0.088653	0.0 0.0 ...	0.000000	0.000000	0.0	0.000000	0.205088	0.000000	0.000000	0.0
2	0.000000	0.046516	0.0 0.000000	0.0 0.0 0.0 0.000000	0.0 0.0 ...	0.076281	0.000000	0.0	0.000000	0.140198	0.000000	0.000000	0.0
3	0.122212	0.000000	0.0 0.246591	0.0 0.0 0.0 0.000000	0.0 0.0 ...	0.000000	0.000000	0.0	0.096160	0.039134	0.000000	0.000000	0.0
4	0.000000	0.000000	0.0 0.000000	0.0 0.0 0.0 0.000000	0.0 0.0 ...	0.000000	0.069083	0.0	0.133335	0.000000	0.000000	0.000000	0.0

5 rows × 1043 columns

We can see that the numerical values corresponding to each token are now float values representing a tf-idf unlike integers we got with the countvectorizer. We also reduced the vocabulary down to 1043 terms, and the vectors are ready to be used for topic modeling, clustering, and classification applications.

Topic modeling

The general aim of topic modeling is uncovering hidden themes or topics in a text corpus. For example, if you had a few thousand documents and knew nothing about them with no labels like in the case of the BBC corpus we are using, then once you have preprocessed and vectorized text, it would be a great idea to take a peek at major topics contained in the corpus.

The output from topic modeling will give you dominant terms or tokens per topic. This will be helpful in naming the topic class itself. For example, if a particular topic had top words such as "football," "baseball," "basketball," and so on, then it would probably be a safe assumption that the topic title should be sports. A second output from the topic model will be the number of topics and their weightages contained within each document. For some algorithms such as latent semantic indexing or analysis (LSI/LSA) and non-negative matrix factorization (NMF), this will just be an absolute number, whereas for latent Dirichlet allocation (LDA), it will be a probability percentage which does add up to one.

It may seem confusing, but each text document encompasses many different "topics"; now, one or two may be dominant, but it's almost impossible to claim that one document belongs in one topic class itself.

However, in real-world problems, we employ this reductionist approach all the time where we assign the dominant topic as the only topic class of the document. This is especially done when we manually label the gold set of text corpus for training supervised classification models.

As a default case, we tend to almost always convert it into a multiclass classification, where one document exclusively belongs to one topic class only, for example, a document may belong to the politics class but not entertainment and vice versa. This is how the BBC corpus is labeled, and for most cases, I recommend that you use this approach since it gives the best bang for the buck with respect to labeling the gold set as well as computational time for training and inference.

This is obviously not the only way to do machine learning classification; we can also perform multilabel classification where each text document can be assigned one or many labels or topics. You can intuitively think of this as running multiple binary classifiers with yes/no response for each class, which may or may not correlate to each other depending on the algorithm and dataset used.

Latent Dirichlet allocation (LDA)

Sklearn contains a module with a well-implemented LDA algorithm so you can directly feed the sparse matrix from the tf-idf vectorizer here as shown in Listing 4-31. There are many important parameters with LDA, but one of the most important is the number of topics.

Listing 4-31. Sklearn's LDA

```
from sklearn.decomposition import LatentDirichletAllocation
num_topics = 4
# for TFIDF DTM
lda_tfidf = LatentDirichletAllocation(n_components=num_topics, random_
state=0)
lda_tfidf.fit(X_train_text)
# Output

LatentDirichletAllocation(batch_size=128, doc_topic_prior=None,
            evaluate_every=-1, learning_decay=0.7,
            learning_method='batch', learning_offset=10.0,
            max_doc_update_iter=100, max_iter=10, mean_change_tol=0.001,
            n_components=4, n_jobs=None, n_topics=None, perp_tol=0.1,
            random_state=0, topic_word_prior=None,
            total_samples=1000000.0, verbose=0)
```

You can query the importance of individual tokens for each topic by querying for .components_ which will be of the shape (number of topics, number of tokens). A much more effective strategy is transposing the dataframe and sorting it so that you can visually see top N tokens per topic as shown in Listing 4-32.

In a true unsupervised learning scenario, you have no prior idea about the text corpus, and hence you will have to start off with some assumption here, get top terms per topic, and see if they make any sense.

I always tell my junior analysts to start off with a topic number equivalent to about 0.25–0.5% of the number of documents and see if the results make any sense; if not, continue to iterate up to 10X of the start number.

Listing 4-32. Printing tokens from top topics

```python
def print_top_words(model, feature_names, n_top_words):
    df_transpose = pd.DataFrame(model.components_, columns = feature_names).transpose()
    topic_names = [col for col in df_transpose.columns]
    for i, topic in enumerate(topic_names):
        message = "Topic #%d: " % topic
        message += " ".join(word for word in list(df_transpose[topic].sort_values(axis = 0, ascending = False).head(n_top_words).index))
        print(message)
    print()
tf_feature_names = np.array(tfidf_transformer.get_feature_names())
print_top_words(lda_tfidf, tf_feature_names, n_top_words = 20)
#output
Topic #0: said mr people year new company government firm market uk 000
sales growth technology mobile 2004 use bank companies world
Topic #1: film mr said labour best blair party election brown awards howard
award mr blair star band actor year minister album prime
Topic #2: game england club win said match players team season play cup
injury time chelsea final ireland year wales world games
Topic #3: chart proposed groups season department prize officials beat
charges poll send increased protect growing create giant japan loss
generation 500
```

From looking at the top 20 tokens for topic 0, it looks like it's combining politics and technology, topic 1 seems to combine entertainment and politics, and so on. So topic delineation for this N_topics isn't all that great, and maybe it's a good idea to run the topic model again with a higher number of topics.

The other important relationship is querying for the percentage of topics for each document. This can be done by simply using the transform method shown in Listing 4-33.

Listing 4-33. Get the percentage of each topic

```
df_transpose = pd.DataFrame(lda_tfidf.components_, columns = tf_feature_
names).transpose()
pd.DataFrame(lda_tfidf.transform(df_dtm.iloc[:1]), columns =
["Topic"+str(col) for col in df_transpose.columns])
# Output
```

	Topic0	Topic1	Topic2	Topic3
0	0.876727	0.041352	0.042201	0.03972

LDA gives you a probability of documents belonging to a particular topic number; hence, it's showing that the dominant topic is topic0.

You can manually check for the quality of this prediction by looking up the text of the original document, or in our case, we can simply cheat a little and look up the label of the document, which is tech in our case, so the prediction is not entirely inaccurate.

We iterate through a list of a number of topics and print top words for each cluster in Listing 4-34. It looks like the topics start overlapping a lot for a number of topics above 6, so the optimum topic number may be 5 or 6.

Listing 4-34. Iterating through different numbers of topics

```
from sklearn.decomposition import LatentDirichletAllocation
num_topics = [5,6,7,8]
def print_lda_terms(num_topics):
    for num_topic in num_topics:
        print("*"*20)
        print("Number of Topics #%d: " % num_topic)
        lda_tfidf = LatentDirichletAllocation(n_components=num_topic,
        random_state=0)
        lda_tfidf.fit(X_train_text)
        print_top_words(lda_tfidf, tf_feature_names, n_top_words = 20)
print_lda_terms(num_topics)
```

```
# Output
********************
```

Number of Topics 5:
Topic #0: said people mr year new company 000 firm market uk music sales growth technology mobile 2004 bank world companies economy
Topic #1: film best awards award actor star films actress oscar comedy singer director stars won tv movie hollywood year series number
Topic #2: game said england club win match play players team time year season cup injury final world chelsea ireland wales old
Topic #3: sign value account growing capital giant non tour limited generation title signed business living finding jobs leader 500 thousands cash
Topic #4: mr said labour blair party election government brown minister mr blair howard prime prime minister mr brown secretary lord tory chancellor leader police

```
********************
```

Number of Topics 6:
Topic #0: said mr people new government labour uk party music election blair told 000 says use make year mobile like technology
Topic #1: film best awards award actor star band album actress singer oscar films comedy won stars rock number director year movie
Topic #2: said growth bank sales oil year economy market company shares 2004 firm prices economic analysts china india profits dollar deal
Topic #3: mr brown mail hands media protect ceremony local straight version cross common ex unless growing launched send reached takes injury current
Topic #4: win final champion match game olympic said world open year time won race second injury cup year old set season old
Topic #5: england club game wales ireland rugby players nations coach squad france team chelsea season league play cup scotland half injury

```
********************
```

Number of Topics 7:
Topic #0: said mr people government new labour music uk party election blair 000 told says use technology mobile year minister make

Topic #1: film best awards award actor star films actress oscar comedy director won movie stars hollywood year tv singer series ceremony
Topic #2: said growth bank sales year market oil company economy shares firm 2004 prices economic analysts china deal india profits dollar
Topic #3: 200 works officials area account digital share price local trial send person protect popular sell bought version growing dollar aimed
Topic #4: 200 works officials area account digital share price local trial send person protect popular sell bought version growing dollar aimed
Topic #5: club chelsea united manchester real shot goal league football post bid contract boss manager premiership area alan champions minutes free
Topic #6: game england win said match players team play cup season injury year final time ireland club world won second coach

****
Number of Topics 8:
Topic #0: said mr people labour government party election blair new uk mobile music technology use minister says software told brown home
Topic #1: film best awards award actor star album band films singer actress oscar comedy won stars director number rock movie hollywood
Topic #2: chelsea club manchester united league champions football real speculation promised department accounts premiership aimed groups bought season demand buying manager
Topic #3: limited dropped williams previously business websites tony figures increase injury rising winning developed suggests irish captain charles giant dollar north
Topic #4: limited dropped williams previously business websites tony figures increase injury rising winning developed suggests irish captain charles giant dollar north
Topic #5: limited dropped williams previously business websites tony figures increase injury rising winning developed suggests irish captain charles giant dollar north
Topic #6: game england said win match club players team season cup play injury final time year ireland world wales coach champion
Topic #7: said company year sales market growth bank firm mr 2004 economy oil shares 000 new economic deal prices china analysts

Interpreting individual topic models gets pretty hard once the number of topics starts going up; and manually going through top 20 terms for each iteration doesn't help much. In such a case, I highly recommend an excellent package called pyLDAvis which helps you interactively visualize top terms per topic based on relevance as well as visually getting the marginal topic distribution in the Jupyter Notebook itself as shown in Figure 4-4. This is a port of the very popular R package called LDAvis; you can check out their detailed methodology in their published paper (`www.aclweb.org/anthology/W14-3110/`). You can not only see top terms per topic, and the overall marginal topic distribution, but also visualize overlap between topics. If we had used a higher number of topics, then the topic circles would have intersected completely. It has been known that pyLDAvis has a bug that consumes lots of memory and may result in errors associated with excessive memory usage when used with certain versions of Python 3.7.x.

Listing 4-35. pyLDAvis example

```
import pyLDAvis
import pyLDAvis.sklearn
# https://github.com/bmabey/pyLDAvis/issues/127
# without sort_topics we will get different topic_ids than what we get
above sklearn offsets start with 0 whereas this starts with 1
num_topics = 5
lda_tfidf = LatentDirichletAllocation(n_components=num_topics, random_
state=0)
lda_tfidf.fit(X_train_text)
pyLDAvis.enable_notebook()
pyLDAvis.sklearn.prepare(lda_tfidf, X_train_text, tfidf_transformer,
mds='mmds', sort_topics=False)
# Output
```

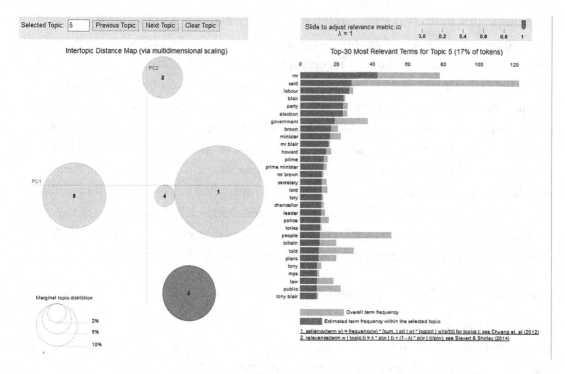

Figure 4-4. *pyLDAvis diagram for sklearn LDA*

Another empirical method to calculate the number of topics is by plotting coherence values against the number of topics. It's been observed that the number of topics near the first couple of maxima values in the coherence plot correlates with the optimum number of topics. This is being increasingly used in practice in the last few years after a couple of interesting papers (`https://dl.acm.org/doi/10.5555/2145432.2145462` and `www.aclweb.org/anthology/D12-1087/`) compared different topic coherence metrics.

We will use the popular Gensim implementation for calculating coherence. It supports different coherence metrics, but we will use umass which is based on the Mimno (2011) paper (`https://dl.acm.org/doi/10.5555/2145432.2145462`).

Let us first convert sklearn tf-idf vectors and apply LDA implementation within Gensim. gensim_corpus is a gensim.matutils.Sparse2Corpus object which is of the same length as the number of documents. gensim_corpus supports indexing so that we can get a list of tuples equal to the number of tokens in our tf-idf_transformer containing gensim_dict index numbers and tf-idf vector weights.

As a sanity check, you can compare the output from gensim_corpus and gensim_dict to the df_dtm created earlier to see if they are the same or not as shown in Listing 4-36.

Listing 4-36. Gensim LDA

```
import gensim
from gensim.corpora.dictionary import Dictionary

def sklearnvect2gensim(vectorizer, dtmatrix):
    corpus_vect_gensim = gensim.matutils.Sparse2Corpus(dtmatrix, documents_
    columns=False)
    dictionary = Dictionary.from_corpus(corpus_vect_gensim,
    id2word=dict((id, word) for word, id in vectorizer.vocabulary_.
    items()))

    return (corpus_vect_gensim, dictionary)

(gensim_corpus, gensim_dict) = sklearnvect2gensim(tfidf_transformer,
X_train_text)
print(type(gensim_corpus))
print(gensim_corpus[0][:10])
print(gensim_dict[gensim_corpus[0][0][0]])
if df_dtm[gensim_dict[gensim_corpus[0][0][0]]].iloc[0] == gensim_corpus[0]
[0][1]:
    print("True")

# Output
<class 'gensim.matutils.Sparse2Corpus'>
[(421, 0.0570566345143319), (145, 0.16739425616154216),
(314, 0.2551493009343519), (167, 0.05427147735416746), (516,
0.23678419419339075), (125, 0.05827332788493036), (943,
0.052998652537136086), (721, 0.10577641952954958), (146,
0.11470305033534169), (341, 0.0821091733728967)]
gordon
True
```

We will also print the top terms in Gensim topics, perplexity, and coherence of the model in Listing 4-37. Gensim uses a slightly different format for displaying top tokens per topic, but the idea remains the same as what we used for printing sklearn top tokens.

Listing 4-37. Exploring topics from Gensim's LDA model

```
lda_model = gensim.models.ldamodel.LdaModel(corpus=gensim_corpus,
                                            id2word=gensim_dict,
                                            num_topics=5,
                                            random_state=100,
                                            update_every=1,
                                            chunksize=10,
                                            passes=10,
                                            alpha='symmetric',
                                            iterations=100,
                                            per_word_topics=True)
def print_gensim_topics(model):
    topics_list = model.print_topics(num_words=20)
    num_topics = len(model.print_topics())
    for i in range(len(topics_list)):
        print("*"*20)
        print("Topics #%d: " % topics_list[i][0])
        print(topics_list[i][1])
print_gensim_topics(lda_model)
from gensim.models import CoherenceModel
print('Perplexity: ', lda_model.log_perplexity(gensim_corpus))

coherence_model_lda = CoherenceModel(model=lda_model,corpus=gensim_corpus,
dictionary=gensim_dict,coherence='u_mass')
coherence_lda = coherence_model_lda.get_coherence()
print('Coherence Score: ', coherence_lda)
vector = lda_model[gensim_corpus[0]]
vector[0]
# Output
********************

Topics #0:
0.012*"said" + 0.010*"uk" + 0.010*"year" + 0.010*"world" + 0.009*"people"
+ 0.008*"new" + 0.008*"number" + 0.007*"mobile" + 0.007*"company" +
0.007*"like" + 0.006*"000" + 0.006*"tv" + 0.006*"phone" + 0.006*"million"
+ 0.006*"high" + 0.006*"home" + 0.006*"dollar" + 0.006*"industry" +
0.005*"market" + 0.005*"used"
```

194

```
********************
Topics #1:
0.017*"2004" + 0.015*"party" + 0.013*"added" + 0.012*"firm" + 0.012*"told"
+ 0.012*"report" + 0.010*"michael" + 0.009*"eu" + 0.009*"country"
+ 0.009*"women" + 0.008*"2003" + 0.008*"office" + 0.008*"net" +
0.008*"looking" + 0.008*"ready" + 0.008*"economic" + 0.008*"issue" +
0.008*"growth" + 0.008*"decision" + 0.008*"london"
********************

Topics #2:
0.042*"film" + 0.040*"game" + 0.029*"best" + 0.029*"games" +
0.027*"players" + 0.026*"play" + 0.018*"champion" + 0.017*"films" +
0.014*"playing" + 0.013*"injury" + 0.013*"open" + 0.013*"actor" +
0.013*"cup" + 0.013*"fans" + 0.013*"award" + 0.013*"awards" + 0.013*"won" +
0.012*"australian" + 0.012*"victory" + 0.012*"movie"
********************

Topics #3:
0.018*"said" + 0.016*"mr" + 0.010*"government" + 0.008*"time" +
0.007*"just" + 0.006*"election" + 0.006*"use" + 0.006*"britain" +
0.006*"tax" + 0.006*"way" + 0.006*"good" + 0.005*"work" + 0.005*"bbc" +
0.005*"old" + 0.005*"help" + 0.005*"set" + 0.005*"brown" + 0.005*"public" +
0.005*"think" + 0.005*"howard"
********************

Topics #4:
0.062*"league" + 0.058*"club" + 0.035*"champions" + 0.029*"manager" +
0.025*"chelsea" + 0.018*"shot" + 0.009*"premiership" + 0.001*"understand"
+ 0.001*"results" + 0.001*"began" + 0.001*"quality" + 0.001*"watch"
+ 0.001*"moved" + 0.001*"ready" + 0.001*"taken" + 0.001*"revealed" +
0.001*"june" + 0.001*"confident" + 0.001*"attack" + 0.001*"charge"
Perplexity:  -7.722395161385292
Coherence Score:  -2.4458274192403513
[(0, 0.019926809),
 (1, 0.08428456),
 (2, 0.7394348),
 (3, 0.12830141),
 (4, 0.0280524)]
```

So the last thing we have to do is plot coherence values in Listing 4-38, and we can easily see a maxima near five topics in Figure 4-5 which is similar to what we observed qualitatively with pyLDAvis as well as by just manually comparing the top tokens and seeing if they seem like they should belong in the same category.

Listing 4-38. Plotting coherence values

```
def calculate_coherence_values(gensim_dict, gensim_corpus, limit, start=2,
step=1):

    coherence_values = []
    model_list = []
    for num_topics in range(start, limit, step):
        lda_model = gensim.models.ldamodel.LdaModel(corpus=gensim_corpus,
                                        id2word=gensim_dict,
                                        num_topics=num_topics,
                                        random_state=100,
                                        update_every=1,
                                        chunksize=10,
                                        passes=10,
                                        alpha='symmetric',
                                        iterations=100,
                                        per_word_topics=True)
        model_list.append(lda_model)
        coherence_model_lda = CoherenceModel(model=lda_model,corpus=gensim_
        corpus, dictionary=gensim_dict,coherence='u_mass')

        coherence_values.append(coherence_model_lda.get_coherence()*-1)

    return model_list, coherence_values
model_list, coherence_values = calculate_coherence_values(gensim_
dict=gensim_dict, gensim_corpus=gensim_corpus, start=2, limit=10, step=1)
import matplotlib.pyplot as plt
limit=10; start=2; step=1;
x = range(start, limit, step)
plt.plot(x, coherence_values)
plt.title("Coherence plot")
```

```
plt.xlabel("Number of topics")
plt.ylabel("Coherence score")
#plt.legend(("coherence_values"), loc='best')
plt.show()
```

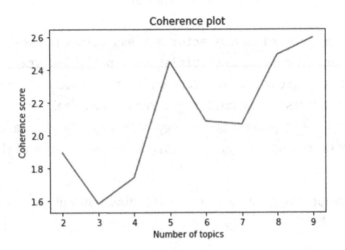

Figure 4-5. *Coherence plot for the LDA topic model*

Non-negative matrix factorization (NMF)

Let us look at NMF for topic modeling in Listing 4-39. This algorithm does not give probabilistic topic values for each document but rather just an absolute number; however, it is usually faster than LDA, and in many cases, this works better than LDA.

Listing 4-39. Sklearn NMF topics

```
from sklearn.decomposition import NMF
n_components = 6
nmf = NMF(n_components=n_components, random_state=1, alpha=.1, l1_ratio=.5)
nmf_tfidf = nmf.fit(X_train_text)
print_top_words(nmf_tfidf, tf_feature_names, n_top_words = 20)
# Output
Topic #0: game win england said match play players cup club team time final
season wales year world injury good chelsea games
```

Topic #1: mr labour blair election party said brown mr blair government minister howard mr brown prime prime minister chancellor tory tax leader tories plans

Topic #2: people music said software users technology microsoft computer digital net internet online new broadband security tv information use web service

Topic #3: film best award awards actor actress director oscar films won star movie comedy prize ceremony british year hollywood stars role

Topic #4: said year growth market economy company sales oil bank 2004 firm economic shares prices china dollar government new deal rise

Topic #5: mobile phone phones technology services use people customers service data networks using video network uk access calls digital devices million

The dominant topic in this case as shown in Listing 4-40 will just be one with the highest absolute value.

Listing 4-40. Dominant NMF topic for a document

```
tf_feature_names = np.array(tfidf_transformer.get_feature_names())
df_transpose = pd.DataFrame(nmf_tfidf.components_, columns = tf_feature_
names).transpose()
pd.DataFrame(nmf_tfidf.transform(df_dtm.iloc[:1]), columns =
["Topic"+str(col) for col in df_transpose.columns])
# Output
```

	Topic0	Topic1	Topic2	Topic3	Topic4	Topic5
0	0.0	0.126342	0.0	0.0	0.061525	0.0

pyLDAvis works on NMF topics too, and we can visualize it in Figure 4-6 by using the same code as Listing 4-35.

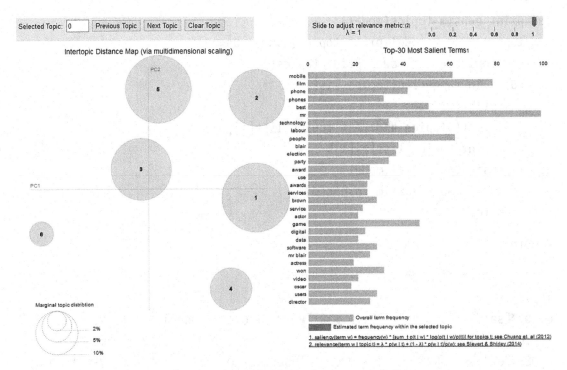

Figure 4-6. *pyLDAvis plot for the NMF-based topic modeling*

Latent semantic indexing (LSI)

Latent semantic indexing, also known as latent semantic analysis (LSA), is based on the dimensionality reduction algorithm known as singular value decomposition (SVD) which is pretty similar to principal component analysis.

Sklearn ships with a truncated SVD algorithm, but we will use Gensim here in Listing 4-41 so that we can directly calculate coherence values for selecting the optimum number of topics. The coefficients of top terms as well as topic values can be negative in LSI, but that doesn't have any physical significance.

Listing 4-41. Gensim's LSI model

```
from gensim.test.utils import common_dictionary, common_corpus
from gensim.models import LsiModel

lsi_model = LsiModel(corpus = gensim_corpus, id2word=gensim_dict,
num_topics=5)
```

```python
def print_gensim_topics(model):
    topics_list = model.print_topics(num_words=20)
    num_topics = len(model.print_topics())
    for i in range(len(topics_list)):
        print("*"*20)
        print("Topics #%d: " % topics_list[i][0])
        print(topics_list[i][1])
print_gensim_topics(lsi_model)
vector = lsi_model[gensim_corpus[0]]
print(vector)

# Output

********************
Topics #0:
0.302*"said" + 0.217*"mr" + 0.138*"year" + 0.132*"people" + 0.121*"new"
+ 0.095*"film" + 0.093*"government" + 0.092*"world" + 0.089*"time" +
0.085*"uk" + 0.083*"game" + 0.082*"labour" + 0.076*"best" + 0.075*"told" +
0.075*"music" + 0.075*"years" + 0.073*"just" + 0.072*"000" + 0.071*"like" +
0.070*"party"
********************
Topics #1:
0.358*"mr" + 0.224*"labour" + 0.188*"election" + 0.187*"blair" +
-0.181*"film" + 0.164*"party" + -0.156*"game" + -0.143*"best" +
0.141*"brown" + 0.141*"government" + 0.133*"mr blair" + 0.116*"minister"
+ 0.115*"mr brown" + 0.106*"tax" + 0.102*"chancellor" + 0.102*"prime" +
0.101*"prime minister" + 0.100*"howard" + -0.100*"win" + -0.095*"play"
********************
Topics #2:
-0.154*"labour" + 0.151*"mobile" + -0.140*"blair" + -0.134*"mr" +
0.126*"market" + -0.124*"election" + 0.122*"firm" + -0.120*"win" +
0.118*"technology" + 0.116*"users" + 0.114*"company" + -0.114*"brown"
+ -0.114*"game" + -0.113*"party" + 0.108*"sales" + 0.108*"phone" +
0.108*"software" + -0.105*"england" + -0.097*"mr blair" + 0.097*"music"
```

```
********************
Topics #3:
-0.570*"film" + -0.271*"best" + -0.186*"award" + -0.180*"awards" +
-0.156*"actor" + 0.147*"game" + -0.142*"actress" + -0.137*"oscar" +
-0.131*"director" + -0.130*"films" + 0.116*"england" + -0.103*"star"
+ -0.092*"movie" + 0.090*"match" + -0.090*"comedy" + 0.089*"club" +
0.088*"players" + 0.087*"cup" + 0.081*"wales" + -0.080*"ceremony"
********************
Topics #4:
-0.203*"people" + 0.179*"economy" + 0.174*"growth" + -0.165*"mobile" +
-0.156*"users" + 0.151*"oil" + -0.145*"software" + -0.145*"technology"
+ 0.141*"economic" + 0.138*"year" + 0.136*"bank" + -0.132*"music" +
-0.124*"phone" + 0.115*"dollar" + 0.115*"prices" + 0.112*"sales" +
-0.111*"microsoft" + 0.109*"shares" + -0.109*"digital" + -0.108*"computer"
[(0, 0.3554863441840997),
 (1, 0.31045356819716763),
 (2, -0.0918747151790444),
 (3, 0.014948992695977962),
 (4, 0.13127935192395496)]
```

We can plot the coherence using a similar function to the LDA model, and here too we notice that the optimum number of topics is five as shown in Figure 4-7 which matches our expectations.

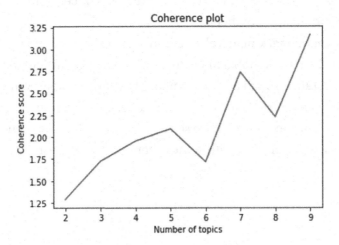

Figure 4-7. *Coherence plot for the LSI topic model*

I provided three major topic modeling algorithms since they are complementary in how they arrive at the optimum number of topics, and sometimes we see that faster models such as LSI give us more intuitive results than LDA, so it doesn't hurt to run all three especially if your dataset is not too large.

Text clustering

In many ways, text clustering is similar to topic modeling since we are still trying to figure out the inherent structure of the corpus without using any labels.

One major difference between the two is that while topic modeling gives us multiple topics per document, text clustering assigns one cluster label to each document within a corpus. So the idea is that a well-tuned clustering algorithm will bin documents into clusters 0, 1, 2, and so on, and all we have to do is have a look at top terms per cluster and simply assign the label, say cluster 0 corresponds to politics, 1 to business, and so on. In an ideal world, text clustering will effortlessly generate a labeled dataset ready for supervised machine learning training for text classification.

We could probably generate labels from topic modeling too by applying a rule-based simplification assigning the topic with the highest weight as the sole topic label for that document. However, it's not always easy to do that if you are dealing with hundreds of topics with quite a few having weights very similar to each other so it's always better to look at results from text clustering algorithms.

Let us look at the kmeans clustering algorithm which aims to bin individual document vectors into a particular cluster with the mean or centroid of the cluster being representative of the cluster members.

Just like topic modeling, kmeans also requires us to specify a hyperparameter value for the number of clusters as shown in Listing 4-42. It is not a particularly fast algorithm, so we should not iterate through a large number of hyperparameter values. The fit_predict method will return an array of cluster labels; we can query for value counts to see the total documents per cluster. If we see a few clusters with very few members, then it may be a good idea to start over with some other number of clusters.

Listing 4-42. Kmeans clustering

```
from sklearn.cluster import KMeans
km = KMeans(n_clusters=8, init='k-means++', n_init=10, max_iter=100,
random_state=0)
#km.fit(X_train_text)

from sklearn.metrics import silhouette_samples
y_km = km.fit_predict(X_train_text)
pd.Series(y_km).value_counts().to_dict()
# Output
{1: 395, 7: 292, 4: 289, 6: 242, 2: 158, 3: 149, 0: 145, 5: 110}
```

We checked top terms per cluster in Listing 4-43 since all the preceding clusters seem to have quite a balanced number of members. We can add cluster numbers as a column to the document term matrix dataframe and filter the dataframe to show documents from individual clusters. Once we have a filtered dataframe, it's just a matter of adding up token weights, transposing it, and sorting it in descending order to display top 30 terms from each cluster.

Listing 4-43. Exploring top terms per cluster

```
df_dtm["cluster_name"] = y_km
df_dtm.head()
cluster_list = len(df_dtm['cluster_name'].unique())
for cluster_number in range(cluster_list):
    print("*"*20)
    print("Cluster %d: " % cluster_number)
    df_cl = df_dtm[df_dtm['cluster_name'] == cluster_number]
    df_cl = df_cl.drop(columns = 'cluster_name')
    print("Total documents in cluster: ", len(df_cl))
    print()
    df_sum = df_cl.agg(['sum'])
    df_sum = df_sum.transpose()
    df_sum_transpose_sort_descending= df_sum.sort_values(by = 'sum',
    ascending = False)
```

```
    df_sum_transpose_sort_descending.index.name = 'words'
    df_sum_transpose_sort_descending.reset_index(inplace=True)
    print(','.join(df_sum_transpose_sort_descending.words.iloc[:30].
    tolist()))
# Output
********************
Cluster 0:
Total documents in cluster:   145

film,best,awards,films,actor,award,oscar,actress,director,star,comedy,year,
won,movie,said,hollywood,stars,ceremony,role,box,british,including,story,
office,tv,new,prize,screen,man,named
********************
Cluster 1:
Total documents in cluster:   395

said,mr,government,music,band,people,uk,new,year,law,000,bbc,police,public,
british,court,lord,told,work,number,ms,years,singer,minister,secretary,
home,children,house,time,rock
********************
Cluster 2:
Total documents in cluster:   158

mr,labour,blair,party,election,said,brown,mr blair,howard,mr
brown,prime,minister,prime minister,government,tory,tax,chancellor,to
ries,leader,campaign,tony blair,tony,britain,people,lib,plans,michael
howard,general election,conservative,public
********************
Cluster 3:
Total documents in cluster:   149

growth,economy,sales,economic,2004,said,prices,year,quarter,rate,rise,
market,figures,dollar,bank,rose,oil,2005,demand,profits,rates,december,
strong,fell,analysts,month,january,jobs,fall,2003
```

```
*******************
Cluster 4:
Total documents in cluster:  289

people,said,mobile,technology,users,software,digital,microsoft,phone,
computer,broadband,net,use,music,games,mr,data,phones,information,internet,
service,new,video,online,using,used,security,mail,web,tv
*******************
Cluster 5:
Total documents in cluster:  110

england,wales,ireland,rugby,france,game,nations,coach,scotland,half,team,
italy,players,squad,injury,captain,win,williams,match,said,saturday,try,
cup,jones,play,andy,centre,victory,ball,international
*******************
Cluster 6:
Total documents in cluster:  242

company,said,mr,firm,shares,market,oil,deal,bank,financial,group,stock,
chief,business,new,state,companies,year,bid,india,government,china,euros,
firms,offer,executive,exchange,investment,investors,analysts
*******************
Cluster 7:
Total documents in cluster:  292

game,said,club,win,play,cup,season,match,year,time,final,world,champion,
team,open,players,old,united,second,good,league,year old,won,olympic,title,
football,set,player,just,goal
```

We can print the top n terms per cluster as before in a much more clean manner if we just work with the .cluster_centers_ attribute as shown in Listing 4-44.

Listing 4-44. Printing top terms using cluster centers

```
len_dict = pd.Series(y_km).value_counts().to_dict()
order_centroids = km.cluster_centers_.argsort()[:, ::-1]
feature_names = tfidf_transformer.get_feature_names()
for i in range(8):
    print("Cluster %d: " % i)
```

```
    print("Total documents in cluster: ",len_dict[i] )
    print()
    temp_list = []
    for word in order_centroids[i, :30]:
        temp_list.append(feature_names[word])
    print(','.join(temp_list))
    print("*"*20)
#Output
Cluster 0:
Total documents in cluster:  145

film,best,awards,films,actor,award,oscar,actress,director,star,comedy,year,
won,movie,said,hollywood,stars,ceremony,role,box,british,including,story,
office,tv,new,prize,screen,man,named
********************

Cluster 1:
Total documents in cluster:  395

said,mr,government,music,band,people,uk,new,year,law,000,bbc,police,public,
british,court,lord,told,work,number,ms,years,singer,minister,secretary,
home,children,house,time,rock
********************

Cluster 2:
Total documents in cluster:  158

mr,labour,blair,party,election,said,brown,mr blair,howard,mr brown,prime,
minister,prime minister,government,tory,tax,chancellor,tories,leader,
campaign,tony blair,tony,britain,people,lib,plans,michael howard,general
election,conservative,public
********************

Cluster 3:
Total documents in cluster:  149

growth,economy,sales,economic,2004,said,prices,year,quarter,rate,rise,
market,figures,dollar,bank,rose,oil,2005,demand,profits,rates,december,
strong,fell,analysts,month,january,jobs,fall,2003
```

```
********************
Cluster 4:
Total documents in cluster:   289

people,said,mobile,technology,users,software,digital,microsoft,phone,
computer,broadband,net,use,music,games,mr,data,phones,information,internet,
service,new,video,online,using,used,security,mail,web,tv
********************
Cluster 5:
Total documents in cluster:   110

england,wales,ireland,rugby,france,game,nations,coach,scotland,half,team,
italy,players,squad,injury,captain,win,williams,match,said,saturday,try,
cup,jones,play,andy,centre,victory,ball,international
********************
Cluster 6:
Total documents in cluster:   242

company,said,mr,firm,shares,market,oil,deal,bank,financial,group,stock,
chief,business,new,state,companies,year,bid,india,government,china,euros,
firms,offer,executive,exchange,investment,investors,analysts
********************
Cluster 7:
Total documents in cluster:   292

game,said,club,win,play,cup,season,match,year,time,final,world,champion,
team,open,players,old,united,second,good,league,year old,won,olympic,title,
football,set,player,just,goal
********************
```

An empirical method to identify the optimum number of clusters is referred to as the "elbow method" where we plot the sum of squared distances of samples to their closest cluster center, known as distortion, vs. the number of clusters. It has been observed that the optimum number of clusters is the point where the slope of the line drastically changes forming a noticeable elbow.

In my experience, the elbow method is a hit or a miss when it comes to detecting the optimum number of clusters for text documents; and as you can see in Listing 4-45 which generates Figure 4-8, there is no major "elbow" but the slope does change at six clusters. When I checked the top terms at six clusters (not shown), it made even less sense than at eight clusters so I suggest we take results from the elbow method with a grain of salt if the elbow is not visually noticeable and always rely on manual checking to see if clusters make qualitative sense.

Listing 4-45. Plotting distortions

```
from sklearn.cluster import KMeans

# checking distortions

distortions = []
for i in range(2, 10):
    #print(i)
    km = KMeans(n_clusters=i, init='k-means++', n_init=10,
    max_iter=100,random_state=0)
    km.fit(X_train_text)
    distortions.append(km.inertia_)
    #print(km.inertia_)
    #print("*"*20)
import numpy as np
from matplotlib import cm
from sklearn.metrics import silhouette_samples
import matplotlib.pyplot as plt

plt.plot(range(2,10), distortions, marker='o')
plt.xlabel('Number of clusters')
plt.ylabel('Distortion')
plt.show()
```

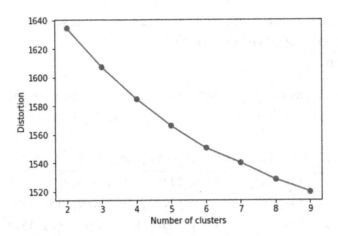

Figure 4-8. *Elbow plot for kmeans clustering*

One drawback with kmeans clustering is its relative slow performance which prevents us from using it on large corpuses.

In those cases, it's worth exploring agglomerative or hierarchical clustering; sklearn implements it with considerable flexibility by allowing a wide variety of distance metrics such as Euclidean, cosine, and so on to compute linkage distances. Since running this algorithm is fast enough, I like to iterate over a list of clusters, as shown in Listing 4-46, and display the number of items per cluster.

Listing 4-46. Agglomerative clustering

```
from sklearn.cluster import AgglomerativeClustering

cluster_list = range(2,10)
def get_optimum_ag_clusters(input_array, cluster_list):
    return_list = []
    for cluster_n in cluster_list:
        temp_dict = {}
        AG = AgglomerativeClustering(n_clusters=cluster_n,
        affinity='euclidean', memory=None, connectivity=None, compute_full_
        tree=True, linkage='ward', pooling_func='deprecated')
        pred_labels = AG.fit_predict(input_array)
        valcount_series = pd.Series(pred_labels).value_counts()
        temp_dict["cluster_n"] = cluster_n
```

```
        temp_dict["cluster_values"] = valcount_series.tolist()
        return_list.append(temp_dict)
    return return_list

return_list = get_optimum_ag_clusters(X_train_text.toarray(), cluster_list)
return_list
#Output
[{'cluster_n': 2, 'cluster_values': [1378, 402]},
 {'cluster_n': 3, 'cluster_values': [1198, 402, 180]},
 {'cluster_n': 4, 'cluster_values': [1042, 402, 180, 156]},
 {'cluster_n': 5, 'cluster_values': [652, 402, 390, 180, 156]},
 {'cluster_n': 6, 'cluster_values': [402, 390, 354, 298, 180, 156]},
 {'cluster_n': 7, 'cluster_values': [390, 354, 298, 291, 180, 156, 111]},
 {'cluster_n': 8, 'cluster_values': [390, 354, 298, 180, 174, 156, 117, 111]},
 {'cluster_n': 9,
  'cluster_values': [390, 354, 298, 174, 156, 134, 117, 111, 46]}]
```

We can shortlist the best ones as shown in Listing 4-47 by simply selecting if the largest cluster has less than 50% of total documents or more than 5% of total documents.

Listing 4-47. Shortlisting optimal number of clusters

```
cluster_labels = []
for items in return_list:
    cluster_labels.append(items["cluster_n"])

for values in return_list:

    sum_value = sum(values["cluster_values"])
    for cluster_num in values["cluster_values"]:
        if 0.05 > cluster_num/sum_value or cluster_num/sum_value > 0.5:
            if values["cluster_n"] in cluster_labels:
                cluster_labels.remove(values["cluster_n"])

print(cluster_labels)
# Output
[5, 7, 8]
```

Once we get a shortlist of the number of clusters, we can run more rule-based elimination strategies which try to see if the size of the largest cluster is going down as the number of clusters increases. Alternatively, we can directly pick the median number of clusters and print the top terms per cluster to see if it makes intuitive sense. We can see that cluster number 7 gives pretty good results as shown in Listing 4-48.

Listing 4-48. Printing top terms per cluster

```
from sklearn.cluster import AgglomerativeClustering

AG = AgglomerativeClustering(n_clusters=7, affinity='euclidean',
memory=None, connectivity=None, compute_full_tree=True, linkage='ward',
pooling_func='deprecated')
pred_labels = AG.fit_predict(X_train_text.toarray())

df_dtm["cluster_name"] = pred_labels
df_dtm.head()

cluster_list = len(df_dtm['cluster_name'].unique())
for cluster_number in range(cluster_list):
    print("*"*20)
    print("Cluster %d: " % cluster_number)
    df_cl = df_dtm[df_dtm['cluster_name'] == cluster_number]
    df_cl = df_cl.drop(columns = 'cluster_name')
    print("Total documents in cluster: ", len(df_cl))
    print()
    df_sum = df_cl.agg(['sum'])
    df_sum = df_sum.transpose()
    df_sum_transpose_sort_descending= df_sum.sort_values(by = 'sum',
    ascending = False)
    df_sum_transpose_sort_descending.index.name = 'words'
    df_sum_transpose_sort_descending.reset_index(inplace=True)
    print(','.join(df_sum_transpose_sort_descending.words.iloc[:30].
    tolist()))
# Output
```

****
Cluster 0:
Total documents in cluster: 291

club,said,win,game,united,champion,cup,year,match,play,season,team,final,
world,open,time,olympic,old,year old,league,players,won,good,second,injury,
set,football,player,just,goal
****
Cluster 1:
Total documents in cluster: 298

people,said,technology,software,mobile,users,games,computer,microsoft,
phone,broadband,digital,use,game,net,mr,video,new,data,online,phones,
security,information,using,service,internet,mail,used,web,content
****
Cluster 2:
Total documents in cluster: 180

mr,labour,election,party,blair,said,brown,mr blair,mr brown,howard,tax,
government,chancellor,minister,prime,prime minister,lord,tory,leader,
people,campaign,tories,tony blair,tony,britain,general,general election,
public,vote,plans
****
Cluster 3:
Total documents in cluster: 156

film,best,award,awards,films,actor,oscar,director,won,actress,year,number,
comedy,said,prize,star,movie,hollywood,british,book,ceremony,stars,
including,role,box,new,named,office,story,uk
****
Cluster 4:
Total documents in cluster: 390

said,company,growth,year,oil,market,firm,bank,mr,economy,sales,shares,2004,
economic,china,prices,government,new,group,analysts,financial,business,000,
chief,rise,stock,quarter,december,2005,state

```
********************
Cluster 5:
Total documents in cluster:  354

said,mr,music,government,band,people,new,year,police,uk,bbc,000,ms,home,
told,public,tv,law,singer,court,british,minister,time,london,plans,
spokesman,years,house,work,men
********************
Cluster 6:
Total documents in cluster:  111

england,wales,ireland,rugby,france,game,nations,scotland,coach,half,team,
players,squad,italy,try,match,captain,win,williams,injury,said,cup,andy,
saturday,play,ball,jones,irish,victory,second
```

Text classification

All the topic modeling and text clustering discussions finally bring us to our ultimate goal of text classification which is the mainstay of processing web scraping data for over a decade.

Text classification is used to automatically determine tags and categories of a particular web page from predetermined categories such as politics, sports, entertainment, and so on.

It is also the technology behind language detection libraries out there such as Python's langid (https://github.com/saffsd/langid.py) where all you do is feed full text from a web page, and it automatically tells you the probability of the natural language it is written.

Such language detection–based filtering is crucial when we are trying to process terabytes of data such as when running a massive web crawler or dealing with large datasets such as common crawl.

Let us take a 40,000 ft view of web scraping; we are trying to extract information from website domains composed of individual web pages, each having text content on a narrow set of topics. Except for a few exceptions, we are trying to extract structured information from a specific subset of web pages, and we need to rely on topic filters based on text classification to filter out all the noise.

One of the most interesting projects at Specrom Analytics was web scraping for some aviation contract–related information from the European Space Agency website (`www.esa.int/`). The client had come to us after being unhappy with the amount of noisy data they got by working with two other consulting companies. We quickly realized the main problem was the varied content on the website; it had an ecommerce section (selling gift shop items), legal and intellectual property (IP), business and contracts, science and technology, and politics, and presumably to increase engagement with general audiences, it also had articles which would frankly be classified as entertainment. We ended up using about 70 different types of binary text classifiers to filter out web pages containing irrelevant content. This case was not typical, but even in our regular web scraping pipeline, we tend to use half a dozen text classifiers to filter out irrelevant information as well as tag and categorize information of interest. Almost all data products from us at Specrom Analytics such as the historical news API which lets you search historical news based on news topics (politics, business, etc.) have a bunch of text classifiers working on the back end.

Sklearn makes training a new text classification model pretty easy; all you need is a vectorized document and a labeled dataset which you should be able to generate now by manually checking the cluster numbers from text clustering or picking the dominant topic.

We will not do much hyperparameter optimization here, but the idea is to iterate through possible hyperparameter combinations by methods such as exhaustive grid searching or random searching and find the combination which maximizes the accuracy of the model. I recommend that you should always optimize your hyperparameters for a production model, and a good starting point is the official sklearn documentation on it (`https://scikit-learn.org/stable/modules/grid_search.html`).

Precision (P), also known as sensitivity, is defined as the number of true positives (Tp) over the number of true positives (Tp) plus the number of false positives (Fp).

$$P = \frac{T_p}{T_p + F_p}$$

Recall (R), also known as specificity, is defined as the number of true positives (Tp) over the number of true positives plus the number of false negatives (Fn).

$$R = \frac{T_p}{T_p + F_n}$$

Instead of measuring precision and recall separately, many pretrained models quote something known as an F1 score which is defined as the harmonic mean of precision and recall.

$$F1 = 2\frac{P \times R}{P + R}$$

We also use classifier accuracy (A) which is defined as a fraction of correct label predictions. Mathematically, it's just a ratio of true positive and true negative over the sum of both true and false positives and negatives.

$$A = \frac{Tp + Tn}{Tp + Tn + Fp + Fn}$$

Sklearn implements all the metrics described here as shown in Listing 4-49.

Listing 4-49. Printing classifier scores

```
from sklearn.metrics import accuracy_score
from sklearn.metrics import recall_score
from sklearn.metrics import precision_score
from sklearn.metrics import f1_score

def print_classifier_scores(train, test, pred_train, pred_test):
    print("Train data accuracy score: ", accuracy_
    score(train["label"],pred_train))
    print("Test data accuracy score: ", accuracy_score(test["label"],
    pred_test))

    print("Recall score on train data: ", recall_score(train["label"],
    pred_train, average='macro'))
    print("Recall score on test data: ", recall_score(test["label"],
    pred_test, average='macro'))

    print("Precision score on train data: ",precision_score(train["label"],
    pred_train, average='macro'))
    print("Precision score on test data: ",precision_score(test["label"],
    pred_test, average='macro'))
```

```
print("F1 score on train data: ", f1_score(train["label"],pred_train,
average='macro'))
print("F1 score on test data: ", f1_score(test["label"],pred_test,
average='macro'))
```

There are numerous supervised learning algorithms we can choose for text classification, but I like to use naive Bayes-based algorithms which scale pretty well on large workloads both for training and inference. We will go through two naive Bayes variants in Listing 4-50.

Listing 4-50. Naive Bayes classifier

```
import time
from sklearn.naive_bayes import MultinomialNB
from sklearn.naive_bayes import ComplementNB

import numpy as np
import pandas as pd
df = pd.read_csv("bbc_news_data.csv")
from sklearn.model_selection import train_test_split
train, test = train_test_split(df, test_size=0.2)
print("Train df shape is: ",train.shape)
print("Test df shape is: ",test.shape)
from sklearn.feature_extraction.text import TfidfVectorizer

tfidf_transformer = TfidfVectorizer(stop_words='english',
                                    ngram_range=(1, 2),max_df=0.97, min_df =
                                    0.03, lowercase=True, max_features=2500)

X_train_text = tfidf_transformer.fit_transform(train['text'])
X_test_text = tfidf_transformer.transform(test["text"])
df_dtm = pd.DataFrame(X_train_text.toarray(), columns=tfidf_transformer.
get_feature_names())

print("Multinomial naive bayes classifier\n")
mnb = MultinomialNB()
train_start_time = time.time()
mnb.fit(X_train_text, train["label"])
train_end_time = time.time()
```

```
print("total time (in milliseconds) to train: ", round(1000*(train_end_
time - train_start_time),3))
pred_train_start_time = time.time()
pred_train = mnb.predict(X_train_text)
pred_train_end_time = time.time()
print("total time (in milliseconds) to predict labels on train data: ",
round(1000*(pred_train_end_time - pred_train_start_time), 3))
pred_test_start_time = time.time()
pred_test = mnb.predict(X_test_text)
pred_test_end_time = time.time()
print("total time (in milliseconds) to predict labels on test data: ",
round(1000*(pred_test_end_time - pred_test_start_time),3))
print_classifier_scores(train, test, pred_train, pred_test)
print("*"*20)

print("Complement Naive Bayes\n")

cnb = ComplementNB(alpha=1.0, fit_prior=True, class_prior=None, norm=False)
train_start_time = time.time()
cnb.fit(X_train_text, train["label"])
train_end_time = time.time()
print("total time (in milliseconds) to train: ", round(1000*(train_end_
time - train_start_time),3))
pred_train_start_time = time.time()
pred_train = cnb.predict(X_train_text)
pred_train_end_time = time.time()
print("total time (in milliseconds) to predict labels on train data: ",
round(1000*(pred_train_end_time - pred_train_start_time), 3))
pred_test_start_time = time.time()
pred_test = cnb.predict(X_test_text)
pred_test_end_time = time.time()
print("total time (in milliseconds) to predict labels on test data: ",
round(1000*(pred_test_end_time - pred_test_start_time),3))
print_classifier_scores(train, test, pred_train, pred_test)
print("*"*20)
#Output
```

Multinomial naive bayes classifier

total time (in milliseconds) to train: 7.02
total time (in milliseconds) to predict labels on train data: 1.002
total time (in milliseconds) to predict labels on test data: 1.055
Train data accuracy score: 0.9707865168539326
Test data accuracy score: 0.9662921348314607
Recall score on train data: 0.9697859913718123
Recall score on test data: 0.9659067173646726
Precision score on train data: 0.9698213452080875
Precision score on test data: 0.9663648248470785
F1 score on train data: 0.9697642938439637
F1 score on test data: 0.96608690414534
****

Complement Naive Bayes

total time (in milliseconds) to train: 6.523
total time (in milliseconds) to predict labels on train data: 1.003
total time (in milliseconds) to predict labels on test data: 1.003
Train data accuracy score: 0.9668539325842697
Test data accuracy score: 0.9640449438202248
Recall score on train data: 0.9654591117837386
Recall score on test data: 0.962606575117162
Precision score on train data: 0.9668281855196993
Precision score on test data: 0.9642158869040557
F1 score on train data: 0.9660095290612176
F1 score on test data: 0.9632802428324068
****

As you can see, we are getting accuracy, precision, and recall hovering at around 96% for training data and unseen or test data. This is pretty good especially considering that it took us only about 6 milliseconds to train the model with 1700 odd documents and just 1 millisecond to make predictions on 445 documents.

We can improve our predictions by using more computationally intensive algorithms such as logistic regression and gradient boosting classifiers or even support vector machines as shown in Listing 4-51 which can be used on small- to medium-sized datasets.

Listing 4-51. Logistic and gradient boosting classifiers

```
from sklearn.linear_model import LogisticRegression
from sklearn.ensemble import GradientBoostingClassifier

print("Logistic Regression\n")
logit = LogisticRegression(solver = 'lbfgs', multi_class = 'auto')

train_start_time = time.time()
logit.fit(X_train_text, train["label"])
train_end_time = time.time()
print("total time (in milliseconds) to train: ", round(1000*(train_end_
time - train_start_time),3))
pred_train_start_time = time.time()
pred_train = logit.predict(X_train_text)
pred_train_end_time = time.time()
print("total time (in milliseconds) to predict labels on train data: ",
round(1000*(pred_train_end_time - pred_train_start_time), 3))
pred_test_start_time = time.time()
pred_test = logit.predict(X_test_text)
pred_test_end_time = time.time()
print("total time (in milliseconds) to predict labels on test data: ",
round(1000*(pred_test_end_time - pred_test_start_time),3))
print_classifier_scores(train, test, pred_train, pred_test)
print("*"*20)

print("Gradient Boosting Classifier\n")
gbc = GradientBoostingClassifier()

train_start_time = time.time()
gbc.fit(X_train_text, train["label"])
train_end_time = time.time()
print("total time (in milliseconds) to train: ", round(1000*(train_end_
time - train_start_time),3))
pred_train_start_time = time.time()
pred_train = gbc.predict(X_train_text)
pred_train_end_time = time.time()
```

```
print("total time (in milliseconds) to predict labels on train data: ",
round(1000*(pred_train_end_time - pred_train_start_time), 3))
pred_test_start_time = time.time()
pred_test = gbc.predict(X_test_text)
pred_test_end_time = time.time()
print("total time (in milliseconds) to predict labels on test data: ",
round(1000*(pred_test_end_time - pred_test_start_time),3))
print_classifier_scores(train, test, pred_train, pred_test)
print("*"*20)
# Output
Logistic Regression

total time (in milliseconds) to train:  168.812
total time (in milliseconds) to predict labels on train data:  1.504
total time (in milliseconds) to predict labels on test data:  1.002
Train data accuracy score:  0.9921348314606742
Test data accuracy score:  0.9707865168539326
Recall score on train data:  0.9922924712103816
Recall score on test data:  0.9693296745020883
Precision score on train data:  0.991755245019527
Precision score on test data:  0.9710673581663698
F1 score on train data:  0.9920061720158057
F1 score on test data:  0.9701367409349061
********************
Gradient Boosting Classifier

total time (in milliseconds) to train:  13858.448
total time (in milliseconds) to predict labels on train data:  17.868
total time (in milliseconds) to predict labels on test data:  5.529
Train data accuracy score:  1.0
Test data accuracy score:  0.9370786516853933
Recall score on train data:  1.0
Recall score on test data:  0.9350859510448137
Precision score on train data:  1.0
Precision score on test data:  0.9378523973234859
```

```
F1 score on train data:  1.0
F1 score on test data:  0.9361976782675707
********************
```

As you can see, it takes orders of magnitude more time to train the model, and the real improvement for unseen data is marginal, ~0.1–1% points. You can definitely squeeze more performance over these baseline models out by tuning hyperparameters and preventing overtraining, but that'll incur even more computational cost. You can also switch to deep learning or neural network–based models which use dense vectors called word embeddings and get that additional 1–3% improvement.

In real terms, after all the tuning and optimization, it means that a gradient boosting–based text classifier may classify 1–3 web pages per 100 more correctly than its equivalent faster naive Bayes type classifier. But it takes orders of magnitude more computational processing to get there.

This may be beneficial if you are trying to win a competition with thousands of dollars as prize money or if your hedge fund client needs a highly accurate sentiments model to classify company-specific news for stock trading.

However, these uses are outliers, and the cost-benefit ratio for obtaining marginal accuracy is entirely an overkill if all you are trying to do is classify documents into English or French so that you do not waste time scraping non-English language pages of a particular website.

Packaging text classification models

Let's say that you trained a classifier model and the performance meets your intended use case. At that point, you should package your trained model for reuse without needing to train it again at inference time. Python's built-in persistence module is pickle, and many people use that for converting models into binaries, but I think joblib does a better job of converting sparse arrays into files.

Let us use joblib to convert our tf-idf vector and complement the naive Bayes model into a binary file in Listing 4-52.

Listing 4-52. Saving a classifier using the joblib library

```
import joblib
joblib.dump(tfidf_transformer, 'tfidfvectorizer.pkl')
joblib.dump(cnb, 'cnbclassifier.pkl')
```

We should always check if the files actually work by converting them back and running them on text data. You can see from Listing 4-53 that the output we get from the metrics is exactly the same as what we got in Listing 4-50, so our pretrained classifier works perfectly.

Listing 4-53. Testing the saved classifier from disk

```
tfidf_pretrained_vectorizer = joblib.load('tfidfvectorizer.pkl')
cnb_pretrained_model = joblib.load('cnbclassifier.pkl')
X_test = tfidf_pretrained_vectorizer.transform(test['text'])
X_train = tfidf_pretrained_vectorizer.transform(train['text'])
pred_test = cnb_pretrained_model.predict(X_test)
pred_train = cnb_pretrained_model.predict(X_train)
print_classifier_scores(train, test, pred_train, pred_test)

# Output

Train data accuracy score:  0.9668539325842697
Test data accuracy score:  0.9640449438202248
Recall score on train data:  0.9654591117837386
Recall score on test data:  0.962606575117162
Precision score on train data:  0.9668281855196993
Precision score on test data:  0.9642158869040557
F1 score on train data:  0.9660095290612176
F1 score on test data:  0.9632802428324068
```

Performance decay of text classifiers

If you use a pretrained text classifier for web scraping for any appreciable period of time, then you may notice that they start going down in performance. This may be baffling since you might have tested, cross-validated, and QA'ed the text classifier on the unseen or test dataset after training it, and it seemed fine at that time. So what's going on?

All text classifiers are trained on the dataset which are from a specific snapshot of time; as time goes by, the content itself starts changing due to the dynamic nature of the Web, and very soon you will find that the training dataset you used to train the text classifier is no longer a good representative of the real-world data you are trying to scrape, and once you realize that, it's no surprise that the performance of most pretrained classifiers starts to drop over time.

Let us take the example of the dataset we worked on in this chapter; it consists of news articles scraped in 2005–2006 from a British news website (BBC). At that time, the political landscape in the UK was dominated by the Labour Party and Tony Blair was the prime minister, so most politics-related articles contained those terms, and hence our tf-idf included that as vectors for our text classification model.

Fast forward to 2020 and the political landscape has changed completely both in the UK and in the world; the UK politics is dominated by Brexit which is one of the top tokens in the BBC politics section in the last few years, and in the world politics section, the top term is Donald Trump, both of which will be completely missed by our text classifier model since the tf-idf model has no tokens corresponding to either term.

We can mitigate this to some extent by using a large training dataset over a longer period of time. We can also loosen our max_df and min_df requirements and take top 50,000–100,000 tokens in tf-idf. This will result in a very sparse high-dimensional array so we can run dimensionality reduction algorithms like SVD to bring it down before training the classifier.

This will not solve the issue completely, and the best recommendation is still to run topic modeling and text clustering algorithms periodically on a fresh dataset to notice a new emerging pattern and retrain the classifier with clustered data.

Some text classifiers such as sentiments, language detection, profanity detection, and NSFW (not safe for work) content detectors will not require retraining much since the underlying tokens powering the classifiers don't really change much over time.

Summary

We learned about how to extract information from plain text using regular expressions and named entity recognition.

We applied a variety of unsupervised learning algorithms to perform topic modeling and cluster text documents to make it easier for us to label the documents.

Lastly, we learned about text classification and packaged a trained model so that we can automatically tag and classify web pages or use them as filters for language detection and so on.

In the next chapter, we will introduce the SQL database and use it to store and query for web-scraped data.

CHAPTER 5

Relational Databases and SQL Language

Relational databases organize data in rows and tables like a printed mail order catalog or a train schedule list and are indispensable for storing structured information from scraped websites.

They are specifically optimized for indexing large amounts of data for quick retrieval. Most relational databases are handled by a database server, an application designed specifically for managing databases and which is responsible for abstracting low-level details of accessing the underlying data.

A relational database management system (RDBMS) is based on the relational model published by Edgar F. Codd of IBM's San Jose Research Laboratory in 1970. Most databases such as SQLite, MySQL, PostgreSQL, Oracle, Microsoft SQL Server, and so on in widespread use are based on the relational database model, but there are few such as Elasticsearch, Cassandra, MongoDB, and so on which don't fit this description and are referred to as NoSQL databases.

Structured Query Language (SQL) is a domain-specific language used in programming and designed for managing data held in an RDBMS or for stream processing in a relational data stream management system (RDSMS).

SQL became a standard of the American National Standards Institute (ANSI) in 1986 and of the International Organization for Standardization (ISO) in 1987; however, not all the features from the standard are implemented in different database systems, and hence SQL code is not completely portable across different RDBMS such as Oracle, MySQL, PostgreSQL, and Microsoft SQL Server.

© Jay M. Patel 2020
J. M. Patel, *Getting Structured Data from the Internet*, https://doi.org/10.1007/978-1-4842-6576-5_5

This chapter will not delve into the fundamentals of the SQL language itself, and readers are encouraged to go through *Sams Teach Yourself SQL in 24 Hours*, 5th ed., by Ryan Stephens, Ron Plew, and Arie D. Jones (Pearson Education, 2011). One advantage of this book is that it starts off with explaining ANSI SQL statements, and once the reader is comfortable with it, they go into minute differences and additional features of major RDBMS implementations.

Another free resource is Khan Academy's introduction to SQL (`www.khanacademy.org/computing/computer-programming/sql`), and you should definitely check it out and brush up on the SQL language before going through this chapter if you are a complete beginner or it's been a while since you last worked on the SQL language.

If you are interested in learning about all the intricacies of database design, then you should check out *Fundamentals of Database Systems*, 7th ed., by Ramez Elmasri and Shamkant B. Navathe (Pearson, 2015). Don't let the name fool you; in the best traditions of computer science textbooks, where most thorough textbooks claim to be a fundamentals or introductions book, this one too is a 1200+ pages behemoth, and it probably answers all the database questions you didn't even know you had before reading this one!

While the RDBMS itself operates on the SQL language, we can use programming language–specific drivers to connect and query the database. The data entered in a SQL database is case sensitive, but the SQL statements are not, and to improve readability and as per convention, they are written in uppercase. Name identifiers such as tables, column names, and so on in lowercase and avoid picking reserved words.

We will cover a lot of ground in this chapter. We will work through a lightweight file-based database called SQLite and connect with it using a GUI editor called DBeaver.

We will also show how to work on the same examples using a full-fledged RDBMS called PostgreSQL if you want to build a production-ready system.

I have included SQLite in this chapter for two reasons. Firstly, it's already packaged with most operating systems so you have no learning curve in trying to download and install it on your computer. Secondly, the SQLite community has been making great efforts to ensure that the expected behavior and syntax of the features such as window functions (`www.sqlite.org/windowfunctions.html`), upserts, and so on mimic those from PostgreSQL. So even if you don't get started with PostgreSQL right now, you will be well placed if you are well versed with SQLite.

We will start off with using SQL statements to access the database using low-level Python-based libraries such as SQLite3 and psycopg2 based on the DB-API2 standard (`www.python.org/dev/peps/pep-0249/`). Once we get a bit more comfortable, we will switch to a higher-level library called SQLAlchemy.

Why do we need a relational database?

Let's better understand use cases for a relational database and its capabilities which make it an attractive choice for data storage for many decades before anyone was even doing web scraping. These types of databases were first conceptualized way back in 1970, and it's become ubiquitous for all the things we interact with in daily life from bank transactions to website back ends to everything in between.

Flat files like CSV are probably what you will think of the most when it comes to storing data in table-like formats. Indeed, such files are incredibly common to transfer data between different pipeline components as well as for general-purpose export and import of data. One obvious advantage for CSV files is the very fast insertion of data since you can quickly create, concatenate, or merge multiple CSV files.

So let's consider a very simple example and use a CSV file as a starting point of the discussion and make our way to a relational system.

Let's say that you are building a clone of Hunter.io such as one we discussed in the regex section in Chapter 4. To refresh your memory, it's an email database website which scrapes email addresses from web pages and allows a user to simply enter a website URL to show a list of all scraped email addresses, URLs of web pages where it found the emails, and date when it scraped it. Let's say that we have a CSV file shown in Figure 5-1 which is very similar to the one generated by Listing 4-3 where we scraped the email addresses from the US FDA warning letters table. We have only added a couple of extra columns to capture the crawl date and the base URL of the email address.

	A	B	C	D	E
1	No.	crawl_date	email	email_base_url	webpage_url
2	0	2020-05-01	Lynn.Bonner@fda.hhs.gov	fda.hhs.gov	https://www.fda.gov/inspections-compliance-enforcement-and-criminal-investigations/warning-letters/nestle-purina-petcare-01022015
3	1	2020-05-01	Lynn.Bonner@fda.hhs.gov	fda.hhs.gov	https://www.fda.gov/inspections-compliance-enforcement-and-criminal-investigations/warning-letters/nestle-purina-petcare-01022015
4	2	2020-05-01	feb@fda.hhs.gov	fda.hhs.gov	https://www.fda.gov/inspections-compliance-enforcement-and-criminal-investigations/warning-letters/sagami-rubber-industries-co-ltd-11242015

Figure 5-1. *CSV file showing scraped email addresses*

In order to replicate the Hunter.io functionality, all we do is search on email_base_url, and we will get all the email addresses we want with crawl_dates and the URLs of where we found them.

Searching through a small CSV file which fits into memory is pretty trivial since all you do is load it up as a pandas dataframe and take it from there. It's absolutely not a dealbreaker anymore even if your CSV file is larger than your server memory since you can do any of the tricks mentioned here (`https://pandas.pydata.org/pandas-docs/stable/user_guide/scale.html`), with the most popular among them being chunking, reading only the subset of columns, and so on.

However, your ability to search through very large CSV files will definitely become inefficient pretty quickly. We need to change our data storage construct if we want reasonably fast query times, and relational databases are one of the most common ways of getting there.

Another issue is the lack of optimum data storage in the preceding file. If you look at the preceding table, then it's clear that there is a lot of data duplication going on, for example, data in rows 1 and 2 both have a duplicate email address and are scraped from the same web page URL. Quite possibly, this could've been prevented by the application which created this CSV file, but still that would've handled a subset of the underlying problem.

What if we had two separate CSV files from two different crawls both having the same email address or web page URL? In that case, the application creating the CSV file would have no idea about the other's existence without iterating through records, and that will make data insertion very slow. It would be great if we can somehow force our system to raise a flag every time we try to insert a row which contains duplicate data in either one column or a combination of columns. Relational databases support "unique values" constraints for precisely solving this problem.

This is not the only type of data duplication going on here; crawl date, crawl name, and email_base_url all seem to have lots of duplicates occupying bytes in our columns, whereas it would've been much better if we could just type out the actual values somewhere else and just reference that in our rows; such references are possible using something called "foreign keys" and are very common in relational databases. This reduces data redundancy, is frequently referred to as "data normalization," and is the bedrock of the relational model.

What is a relational database?

The basic unit of a relational database is a table, with rows and columns as you would see in any spreadsheet.

The difference is that the data which can go into the columns is defined beforehand and is known as a database schema.

We will explain how to implement a database schema in the next section, but I just want you to understand for now that a flat table in a CSV file is broken down into multiple tables in a relational database.

Relational databases do incur a size overhead compared to a CSV file due to provisioning for all the features described here; overall though storage requirements are still reduced due to data normalization.

In real life though, we need to balance the need for minimizing disk storage via data normalization, quick inserts, and fast lookup query times since all of these inherently work in opposite. A fully normalized database will require a lot of lookups on foreign keys (called "joins"), and this will increase querying times.

Databases which are intended to perform a large number of transaction type queries (inserting, updating, deleting data), such as those performed by a bank or credit card company, are known as online transactional processing (OLTP) systems. Generally speaking, these databases are highly normalized, and we measure the performance of such databases based on how many transactions they can handle per second.

On the other hand, we have data warehouse systems known as online analytical processing (OLAP) which are intended to perform complex aggregational queries based on selecting data for business intelligence, and since query times here can run into several minutes, quite frequently the data is denormalized to allow faster query times. Examples of common schema in this class are star, snowflake, and galaxy schema.

We also have hybrid transactional/analytical systems which combine the attributes of both the models earlier.

In reality, a database schema design is a pretty vast topic especially when you consider NoSQL alternatives, but I just wanted to emphasize that data normalization has to match your individual use case. If you follow conventional guidelines and normalize to an extreme, then you may end up with a grossly inefficient system with ridiculously long query times.

The major attributes of relational database transactions are known as ACID and are described as follows:

- **Atomicity**: Database modifications must follow an all or nothing rule where the entire transaction fails even if one part of it fails. A database management system should maintain the atomic nature of transactions in spite of any DBMS, operating system, or hardware failure.

- **Consistency**: Database modifications should allow only valid data to be written to the database, and all the changes made by a transaction must leave the database in a valid state as defined by any constraints and other rules. If a transaction is executed that violates the database's consistency rules, the entire transaction will be rolled back, and the database will be restored to a state consistent with those rules.

- **Isolation**: Multiple transactions occurring at the same time do not impact each other's execution.

- **Durability**: Any transaction committed to the database will not be lost or rolled back.

Not all relational database engines are ACID compliant; however, for our purposes, we will only work with ACID-compliant SQL databases in this chapter.

Data definition language (DDL)

SQL statements which deal with the creation or modification of database objects such as tables, views, and index are referred to as data definition language (DDL); basically, any manipulation to the database schema itself is performed using DDL.

A **primary key constraint** uniquely identifies each row in a table, and any column with unique data can be set as primary keys such as columns containing social security numbers.

Data in multiple tables are all linked together by a **foreign key** which is a column in a child table that references a primary key in the parent table. This constraint helps cross-referenced data consistent across tables.

The general SQL syntax for DDL is CREATE TABLE followed by column names and data types in parentheses.

There are different data types available you can take a look at the official documentation for SQLite and PostgreSQL for more details since those are what we will be using in this chapter, but all RDBMS systems have their own flavor of data types in addition to normally available ones.

Primary and foreign keys are explicitly mentioned, and we can also designate a unique constraint on individual columns or groups of columns. The general DDL statement is shown as follows:

```
CREATE TABLE table_name_1 (
    column_name_1_id datatype,
    column_name_2_id datatype,
    column_name_3 datatype,
    column_name_4 datatype,
    .
    .
```

```
    PRIMARY KEY (column_name_1_id),
    CONSTRAINT unique_constraint_name UNIQUE (column_name_3, column_name_4),
    FOREIGN KEY(column_name_2_id) REFERENCES table_name_2 (column_name_1_id)
);
```

DROP will delete entire tables, views, or index. You can alter a table by the ALTER command with a similar syntax. Finally, TRUNCATE can be used to delete all the data within a table, but not the table itself. This is similar to the DELETE clause in the next section; however, truncate will delete everything without any ability to select individual rows.

```
DROP object_type object_name;
ALTER object_type object_name;
TRUNCATE TABLE table_name_1;
```

You should periodically run vacuum to reclaim disk memory for deleted or dropped tables. This can be done by simply typing the VACUUM command if you want to perform it on the entire database.

```
VACUUM;
```

Sample database schema for web scraping

Let's say that we want to store scraped data from web pages in a database. It is supposed to power an email database API so that someone can enter a domain address and can easily find all email addresses associated with that domain as well as a list of all URLs and dates of when those pages were crawled. Apart from that, we should also store other pertinent information from the web crawls which can serve as a back-end database to a broader set of data APIs rather than just emails.

We typically would also want to store contents of the web pages themselves, such as full text, title, primary image, date, author name, and so on.

We can do all of the preceding things with the help of individual tables. Our chief goal when designing a database schema composed of individual columns and tables is achieving a happy optimum between the normalization of data aka minimum amount of duplication data between tables, fast query times, and quick insertion times and minimizing storage requirements.

- All documents are scraped as part of a bulk crawl job, so let's first create a crawl_table which has a crawl_id (primary key) and crawl_url which includes the location of where the raw crawl files are located; this could be a local filesystem, an S3 directory, and a timestamp column which includes the crawl date. I like to create crawl segment files which are decently sized such as 800 MB–1 GB so that it can comfortably fit in memory of the server to do further processing.

- Every web page we crawl belongs to a root domain which is unique. Let's store information pertaining to the root domain in a table called "sources." For example, a web page jaympatel.com/about belongs to a unique root domain called jaympatel.com, and the .com part is called a "top-level domain"; if it was a country extension such as ".co.uk", then we would say that the root domain was registered at the country code second-level domain. Let's assign a unique id called source_id, and this is the primary key. Other columns include source_url. We should also have a column for describing the source. For example, if our source_url was `https://www.fda.gov`, then its description could simply be the official US governmental agency in charge of food safety, cosmetics, pharmaceuticals, and medical devices.

- Every web page we crawl will have a unique URL. We should have a table which contains a web page id, web page URL, crawl id, and source_id.

- Full text, title, and so on from all web pages can be stored in a separate table called documents. It will have a unique document id which is its primary key, a crawl_id which is its foreign key from the crawl table, and a timestamp column for the date of the document. Note that the date here is parsed directly from the document itself, and this is usually different from the crawl_date found in the crawl table. We will also have a topic column which can be filled by a text classification model. It's also possible to assign information in the topics column using a rule based or heuristics, pattern matching or regular expressions or using topic classification–based machine learning models. For example, when we were scraping the warning letters table of the US FDA, we know that all the links point to the warning letters themselves, so we can safely assign all the documents coming out of that crawl with a "warning letter" tag. Alternatively, if we observe the URL patterns of the documents themselves, then all of them contain "inspections-compliance-enforcement-and-criminal-investigations/warning-letters" which is another way for us to autoassign the topic keyword.

- All web pages have certain assets we like to crawl; let's say you want to extract out all email addresses found on a web page. Let's call this table emails. The columns will be email_id, extracted email address, and source_id to point to the domain of the email address itself.

- We have to link email_id (foreign key from the email addresses table) with webpage_id (from the webpage table) where we found the email. Note that the domain where you find an email address may not be the same as the source id which the email address itself belongs to. If the same email_id is found on multiple webpage_id, then it could be a signal that the email address actually exists; however, if we happened to find a large number of email addresses on a particular webpage_id, then it could be a negative signal that the original page could be a spam list and the domain address itself may be a suspect. We can use this information to guide the frequency of our crawler to index a particular spammy page and instead focus our computational power on indexing information from legitimate sites.

- We can round off our database schema by including a table called persons and article_authors which can capture authorship information for a web page if it happens to be a blog, news, or other types of a text document.

In Figure 5-2, each table is depicted by a box, and names of columns are written inside it. Unique ids which represent the individual rows are in bold and as a convention appended with "id" at the end. Relations between tables in the form of foreign keys are represented by dashed lines.

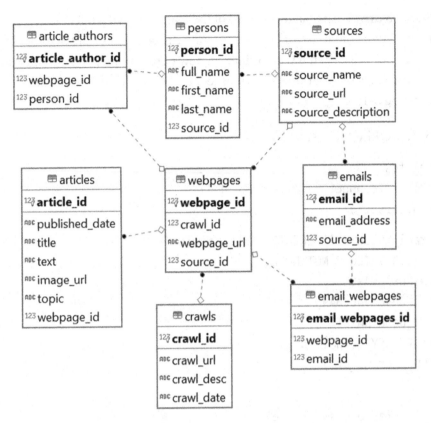

Figure 5-2. *Database schema for storing structured information from web scraping*

SQLite

SQLite is a lightweight RDBMS system which is packaged as a C library, and databases in SQLite work are simply embedded as files on host systems. It provides the best of both worlds, convenience of filesystems as well as powerful SQL language interface and relational data model.

Almost all major operating systems package the SQLite library, and it is also found embedded on a wide variety of other computing platforms and consumer electronics such as mobile phones and so on.

You can initialize a new SQLite database file by simply typing SQLite3 on your bash or command-line console. It will initialize an in-memory database; you can type ".save databasename" to save it on your filesystem.

The SQLite3 driver is packaged as part of the standard Python 3 library, and we can create a new SQLite3 database in Python by using the connect of the SQLite3 library. If a database does not exist, then it will initialize a new database file. You can pass any SQL statements as strings by creating a cursor object and using the execute statements as shown in Listing 5-1.

Listing 5-1. SQLite DDL statements

```
import sqlite3
conn = sqlite3.connect("sqlite-test.db")
cur = conn.cursor()

create_crawl_table = '''CREATE TABLE crawls (
    crawl_id INTEGER NOT NULL,
    crawl_url VARCHAR,
    crawl_desc VARCHAR,
    crawl_date DATETIME,
    PRIMARY KEY (crawl_id),
    UNIQUE (crawl_url)
);'''

cur.execute(create_crawl_table)

create_sources_table = '''CREATE TABLE sources (
    source_id INTEGER NOT NULL,
    source_name VARCHAR,
    source_url VARCHAR,
    source_description VARCHAR,
    PRIMARY KEY (source_id),
    UNIQUE (source_name),
    UNIQUE (source_url)
);'''

cur.execute(create_sources_table)

create_webpages_table = '''CREATE TABLE webpages (
    webpage_id INTEGER NOT NULL,
    crawl_id INTEGER,
    webpage_url VARCHAR,
```

```
    source_id INTEGER,
    PRIMARY KEY (webpage_id),
    CONSTRAINT unique_webpage_crawl UNIQUE (webpage_url, crawl_id),
    FOREIGN KEY(crawl_id) REFERENCES crawls (crawl_id),
    FOREIGN KEY(source_id) REFERENCES sources (source_id)
);'''

cur.execute(create_webpages_table)

create_emails_table = '''CREATE TABLE emails (
    email_id INTEGER NOT NULL,
    email_address VARCHAR,
    source_id INTEGER,
    PRIMARY KEY (email_id),
    UNIQUE (email_address),
    FOREIGN KEY(source_id) REFERENCES sources (source_id)
);'''

cur.execute(create_emails_table)

create_persons_table = '''CREATE TABLE persons (
    person_id INTEGER NOT NULL,
    full_name VARCHAR,
    first_name VARCHAR,
    last_name VARCHAR,
    source_id INTEGER,
    PRIMARY KEY (person_id),
    UNIQUE (full_name),
    FOREIGN KEY(source_id) REFERENCES sources (source_id)
);'''

cur.execute(create_persons_table)

create_email_webpages_table = '''CREATE TABLE email_webpages (
    email_webpages_id INTEGER NOT NULL,
    webpage_id INTEGER,
    email_id INTEGER,
    PRIMARY KEY (email_webpages_id),
    CONSTRAINT unique_webpage_email UNIQUE (webpage_id, email_id),
```

237

```
    FOREIGN KEY(webpage_id) REFERENCES webpages (webpage_id),
    FOREIGN KEY(email_id) REFERENCES emails (email_id)
);'''

cur.execute(create_email_webpages_table)

create_articles_table = '''CREATE TABLE articles (
    article_id INTEGER NOT NULL,
    published_date DATETIME,
    title VARCHAR,
    text VARCHAR,
    image_url VARCHAR,
    topic VARCHAR,
    webpage_id INTEGER,
    PRIMARY KEY (article_id),
    FOREIGN KEY(webpage_id) REFERENCES webpages (webpage_id)
);'''

cur.execute(create_articles_table)

create_article_authors_table = '''CREATE TABLE article_authors (
    article_author_id INTEGER NOT NULL,
    webpage_id INTEGER,
    person_id INTEGER,
    PRIMARY KEY (article_author_id),
    CONSTRAINT unique_article_authors UNIQUE (webpage_id, person_id),
    FOREIGN KEY(webpage_id) REFERENCES webpages (webpage_id),
    FOREIGN KEY(person_id) REFERENCES persons (person_id)
);'''

cur.execute(create_article_authors_table)
conn.commit()
cur.close()
conn.close()
```

DDL statements commit automatically, but it is a good practice to explicitly commit all your changes and close the connection object if you don't need it anymore.

DBeaver

DBeaver (`https://dbeaver.io/`) is a graphical user interface (GUI)–based database client application with compatibility across many RDBMS systems. It helps us visualize the database schema and also lets us connect to a wide variety of databases such as SQLite, PostgreSQL, and so on. You can connect to a database by clicking database ➤ New database connection from the top menu, and it opens up a connection wizard as shown in Figure 5-3. Just select the appropriate database type and click next.

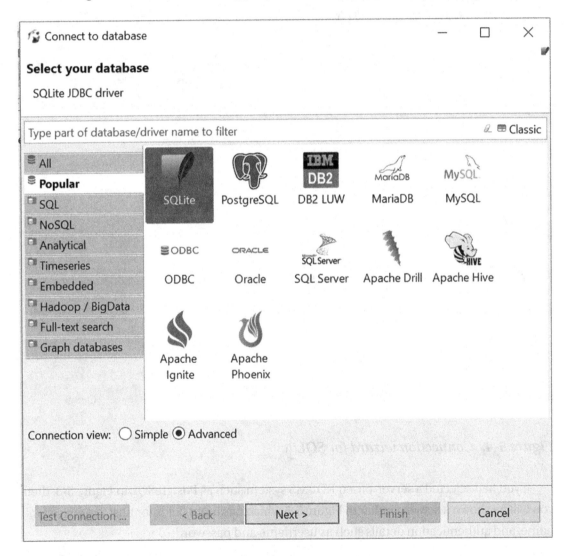

Figure 5-3. *Connection wizard on DBeaver*

For SQLite, you can simply enter the file path of the SQLite database file on your local system in the path text box as shown in Figure 5-4.

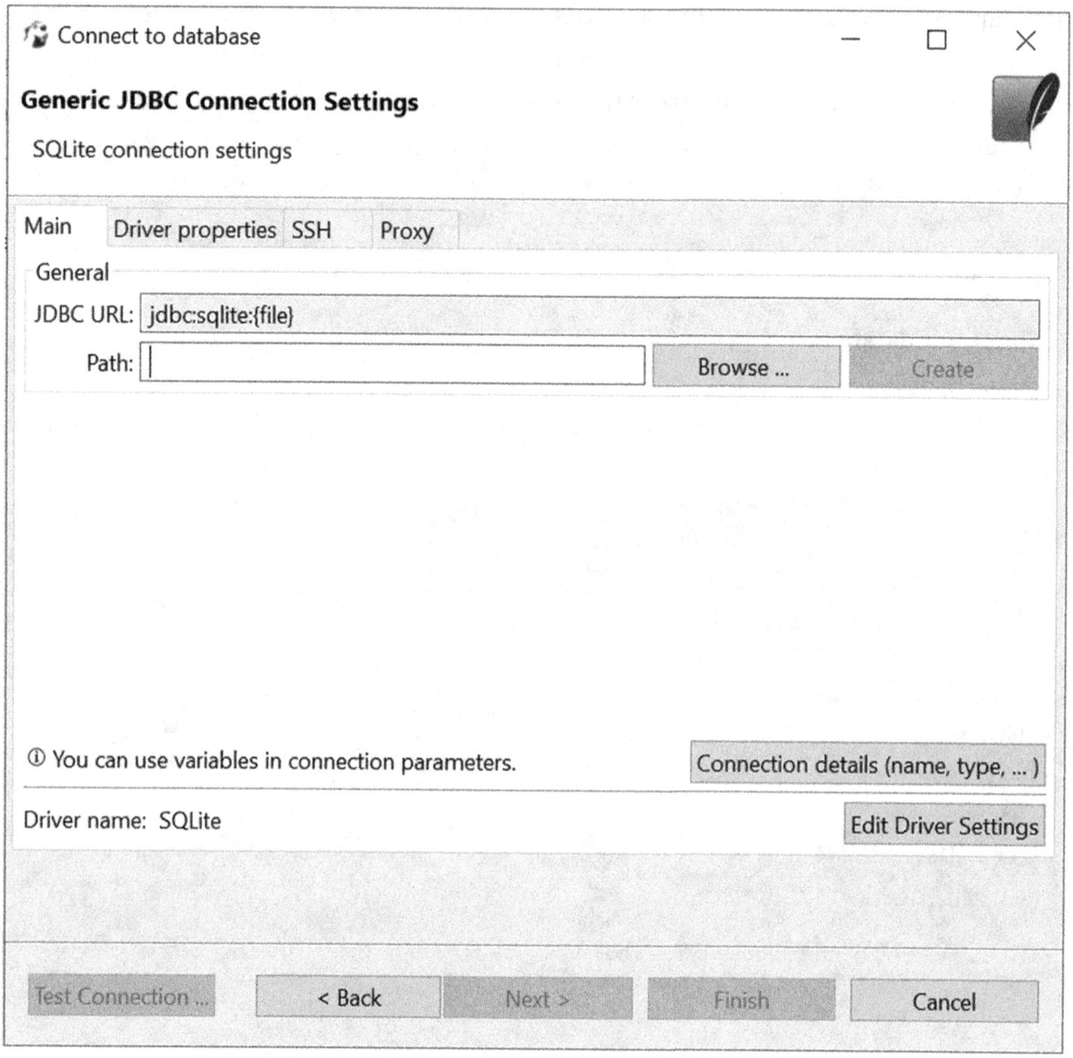

Figure 5-4. *Connection wizard for SQLite*

If you had selected a server-based RDBMS system such as PostgreSQL in Figure 5-3, then you will be shown a screen similar to Figure 5-5 which has input fields for host, database name, and authentication details such as username and password.

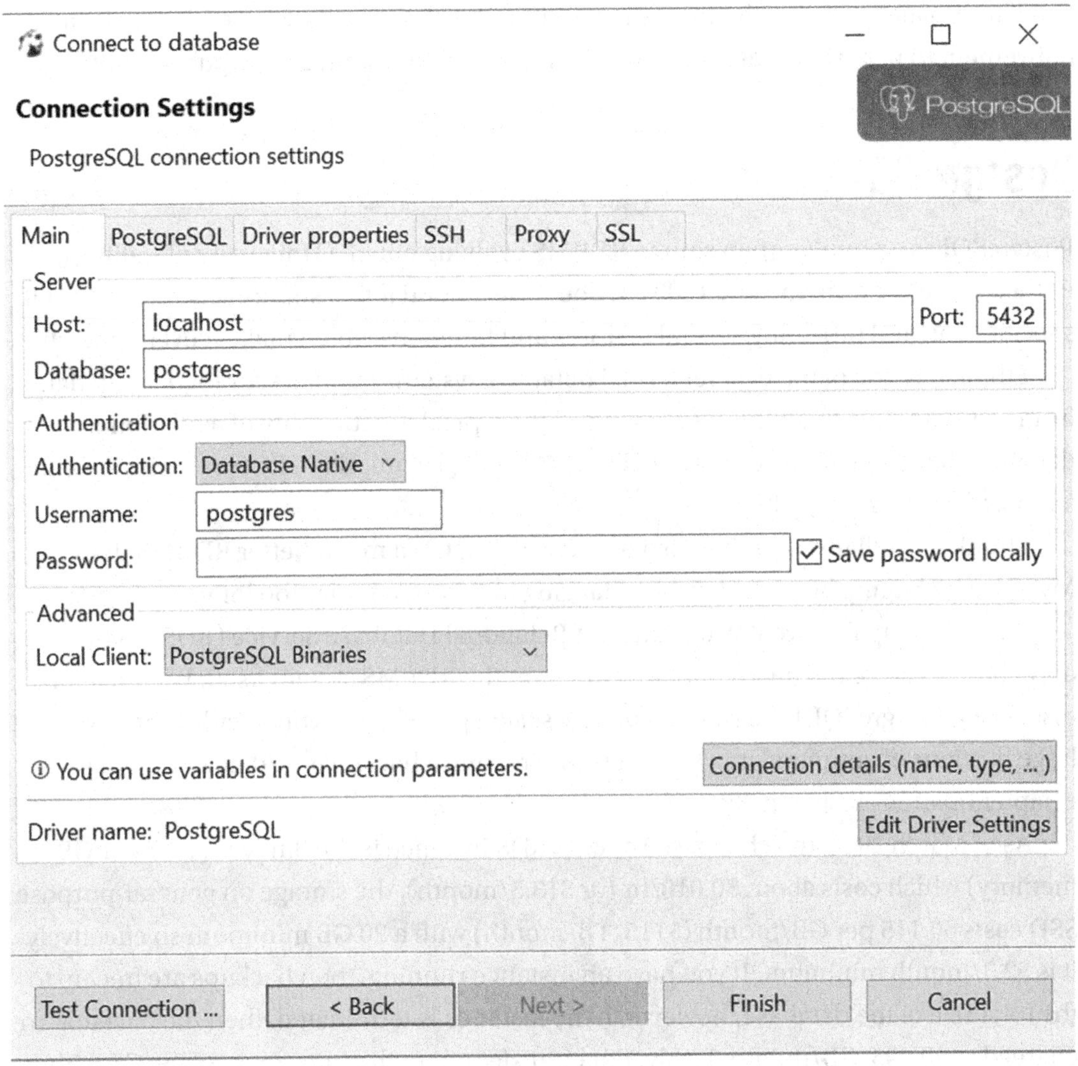

Figure 5-5. *Connection page for PostgreSQL*

Once a new database is connected in DBeaver, you can visualize the tables by using the left explorer pane, and you can get more details as well as a schema diagram by clicking the ER diagram on the right pane.

DBeaver also allows you to directly query the database by simply opening a SQL editor by right-clicking the database of interest.

We have used raw SQL queries in Listing 5-1 so you can paste the create table queries into a DBeaver SQL editor and create the tables that way.

This is usually very useful for testing and troubleshooting; however, it's usually more convenient to pass DDL statements in a higher-level programming language itself.

PostgreSQL

PostgreSQL is a popular open source RDBMS evolving out of a University of California, Berkeley, project from the late 1980s. It does a very good job of supporting the ANSI SQL specification and providing several features and functions which make it even powerful.

I think it is one of the best relational database systems out there for loading up data from web scraping. In addition to having all the typical functionality of a relational database, being open source and ACID and ANSI SQL compliant, it comes with specific support for full text searching.

I think this full text search is also why PostgreSQL is a much better RDBMS than MySQL which quite honestly has been lagging in this area for far too long.

We will set up PostgreSQL on Amazon Relational Database Service (RDS) so that we don't have to delve too much into how to set up the RDBMS system itself. We are mainly discussing PostgreSQL here so that you can scale up to 64 TB if you decide to index large portions of common crawl datasets for creating a backlinks and news database in Chapters 6 and 7, respectively.

As of March 2020, the cheapest Amazon RDS instance is db.t3.micro (1 core, 1 GB memory) which costs about $0.018/hr (or $13.5/month); the storage on general-purpose SSD costs $0.115 per GB/month ($115/TB/month) with a 20 GB minimum so effectively it is $2.3/month minimum. If you have an instance running, then backups are free up to the total size of the database; however, if the instance is terminated, then the backups are charged at $0.095/GB/month. Lastly, data transfer out to the Internet is $0.09/GB with first 1 GB free.

On the other hand, the cheapest Amazon Lightsail database instance costs $15/month, and it includes 1 core, 1 GB memory, 40 GB SSD storage, and 100 GB data transfer, and backups are $0.05/GB/month. An additional storage is available for $0.1/GB/month in 8 GB increments. If you are creating a small production app, then it might make more sense to check out the Lightsail based managed database, but in this chapter, we will set up an RDS instance since that will allow for your database to scale to very high workloads.

In the interest of being concise, we will not get into how to assign fine-grained permissions to the database tables in production, but it is routinely done on RDBMS systems, and SQL statements that do that are known as data control language (DCL) statements. I will encourage you to check out the official AWS RDS documentation on that subject (`https://aws.amazon.com/blogs/database/managing-postgresql-users-and-roles/`).

Setting up AWS RDS PostgreSQL

Go to AWS RDS console (`https://console.aws.amazon.com/rds/home`), go to create database, and go through the options.

1. Pick standard create for the database creation method.

2. Click PostgreSQL in engine options.

3. Pick a version 11.x, preferably 11.6-R1.

4. Select "free tier" for templates.

5. Set up the database name and credentials of your choice.

6. Keep everything else as default and click create database at the end of the page.

7. After the db is created, click the VPC security group link shown in Figure 5-6.

Connectivity & security

Endpoint & port	Networking	Security
Endpoint database-2.cnjff0fgfmxw.us-east-2.rds.amazonaws.com	Availability zone us-east-2a	VPC security groups default (sg-c54aafa9) (active)
Port 5432	VPC vpc-e12e1489	Public accessibility Yes
	Subnet group default-vpc-e12e1489	Certificate authority rds-ca-2019
	Subnets subnet-d6e9cbbe subnet-b8bc7ff4 subnet-da6bf3a0	Certificate authority date Aug 22nd, 2024

Figure 5-6. *VPC security group settings*

Once you are at the security group page, click the inbound rule, click edit, and make sure you open the port 5432 for PostgreSQL as in Figure 5-7.

Type ⓘ	Protocol ⓘ	Port Range ⓘ	Source ⓘ	Description ⓘ
HTTP	TCP	80	0.0.0.0/0	
HTTP	TCP	80	::/0	
PostgreSQL	TCP	5432	0.0.0.0/0	
PostgreSQL	TCP	5432	::/0	
MYSQL/Aurora	TCP	3306	0.0.0.0/0	
MYSQL/Aurora	TCP	3306	::/0	

Figure 5-7. *Inbound rules for the database*

We can use a popular Python library called psycopg2 to communicate with the PostgreSQL server. It is compliant with the DB-API2 standard so the library itself is pretty similar to the usage of the SQLite3 library we saw in the last section.

The wrinkle is in minute differences in SQL supported by PostgreSQL and SQLite which makes it necessary for us to make edits to the DDL statements in Listing 5-2. The major difference is that PostgreSQL uses a serial data type for primary keys. The other difference is that for dates, PostgreSQL has a data type called "timestamp without time zone" vs. no separate date data type for SQLite which makes it necessary to use a built-in function called datetime for the same purpose.

Listing 5-2. PostgreSQL DDL statements

```python
import psycopg2
conn = psycopg2.connect(dbname="database_name", user="user_name",
password="password", host= "enter_database_url_here")
cur = conn.cursor()

create_crawl_table = '''CREATE TABLE crawls (
    crawl_id SERIAL NOT NULL,
    crawl_url VARCHAR,
    crawl_desc VARCHAR,
    crawl_date TIMESTAMP WITHOUT TIME ZONE,
    PRIMARY KEY (crawl_id),
    UNIQUE (crawl_url)
);'''
```

```
cur.execute(create_crawl_table)

create_sources_table = '''CREATE TABLE sources (
    source_id SERIAL NOT NULL,
    source_name VARCHAR,
    source_url VARCHAR,
    source_description VARCHAR,
    PRIMARY KEY (source_id),
    UNIQUE (source_name),
    UNIQUE (source_url)
);'''

cur.execute(create_sources_table)

create_webpages_table = '''CREATE TABLE webpages (
    webpage_id SERIAL NOT NULL,
    crawl_id INTEGER,
    webpage_url VARCHAR,
    source_id INTEGER,
    PRIMARY KEY (webpage_id),
    CONSTRAINT unique_webpage_crawl UNIQUE (webpage_url, crawl_id),
    FOREIGN KEY(crawl_id) REFERENCES crawls (crawl_id),
    FOREIGN KEY(source_id) REFERENCES sources (source_id)
);'''

cur.execute(create_webpages_table)

create_emails_table = '''CREATE TABLE emails (
    email_id SERIAL NOT NULL,
    email_address VARCHAR,
    source_id INTEGER,
    PRIMARY KEY (email_id),
    UNIQUE (email_address),
    FOREIGN KEY(source_id) REFERENCES sources (source_id)
);'''
```

```
cur.execute(create_emails_table)

create_persons_table = '''CREATE TABLE persons (
    person_id SERIAL NOT NULL,
    full_name VARCHAR,
    first_name VARCHAR,
    last_name VARCHAR,
    source_id INTEGER,
    PRIMARY KEY (person_id),
    UNIQUE (full_name),
    FOREIGN KEY(source_id) REFERENCES sources (source_id)
);'''

cur.execute(create_persons_table)

create_email_webpages_table = '''CREATE TABLE email_webpages (
    email_webpages_id SERIAL NOT NULL,
    webpage_id INTEGER,
    email_id INTEGER,
    PRIMARY KEY (email_webpages_id),
    CONSTRAINT unique_webpage_email UNIQUE (webpage_id, email_id),
    FOREIGN KEY(webpage_id) REFERENCES webpages (webpage_id),
    FOREIGN KEY(email_id) REFERENCES emails (email_id)
);'''

cur.execute(create_email_webpages_table)

create_articles_table = '''CREATE TABLE articles (
    article_id INTEGER NOT NULL,
    published_date DATETIME,
    title VARCHAR,
    text VARCHAR,
    image_url VARCHAR,
    topic VARCHAR,
    webpage_id INTEGER,
    PRIMARY KEY (article_id),
    FOREIGN KEY(webpage_id) REFERENCES webpages (webpage_id)
);'''
```

```
cur.execute(create_articles_table)

create_article_authors_table = '''CREATE TABLE article_authors (
    article_author_id SERIAL NOT NULL,
    webpage_id INTEGER,
    person_id INTEGER,
    PRIMARY KEY (article_author_id),
    CONSTRAINT unique_article_authors UNIQUE (webpage_id, person_id),
    FOREIGN KEY(webpage_id) REFERENCES webpages (webpage_id),
    FOREIGN KEY(person_id) REFERENCES persons (person_id)
);'''

cur.execute(create_article_authors_table)

conn.commit()
cur.close()
conn.close()
```

SQLAlchemy

SQLAlchemy is an object-relational mapper (ORM) and a SQL toolkit in Python which abstracts away the minute differences in SQL language implementations across major RDBMS systems and allows us to write relatively portable SQL code from within Python itself.

It consists of SQLAlchemy Core consisting of the SQL expression language which is a thin wrapper around the SQL language itself and an ORM module which is a much higher-level abstraction built on top of the SQL expression language.

We can use SQLAlchemy to write DDL statements and select the database engine of our choice. Based on the engine selected, it will automatically modify our SQL statements to conform to the database of our choice. You can look at the raw SQL queries passed to the database by SQLAlchemy by setting the echo parameter in the create_ engine to true.

We can create an engine for SQLite or PostgreSQL using Listing 5-3 or Listing 5-4, respectively.

Listing 5-3. SQLAlchemy engine for SQLite

```
from sqlalchemy import create_engine
engine = create_engine(r'sqlite:///sqlite-test.db', echo=True)
```

Listing 5-4. SQLAlchemy engine for PostgreSQL

```
from sqlalchemy import create_engine

db_string = "postgresql+psycopg2://username:password3@enter_database_url_
here:5432/databasename"
engine = create_engine(db_string,echo=True)
```

Once you have connected to the database, we should use the reflect method shown in Listing 5-5 to check for tables already created.

Listing 5-5. Table reflection

```
from sqlalchemy.engine import reflection
insp = reflection.Inspector.from_engine(engine)
print(insp.get_table_names())

# Output
2020-06-23 10:32:21,223 INFO sqlalchemy.engine.base.Engine SELECT name FROM
sqlite_master WHERE type='table' ORDER BY name
2020-06-23 10:32:21,224 INFO sqlalchemy.engine.base.Engine ()
['article_authors', 'articles', 'crawls', 'email_webpages', 'emails',
'persons', 'sources', 'webpages']
```

If it shows that there are existing tables, then it will be better if we populate our metadata object and drop all tables in metadata as shown in Listing 5-6 before proceeding further.

Listing 5-6. SQLAlchemy metadata

```
from sqlalchemy import MetaData
metadata = MetaData()
metadata.reflect(engine)
metadata.drop_all(engine)
```

SQLAlchemy's DDL statements shown in Listing 5-7 are pretty intuitive to understand so we won't get too much into details, but since we have turned on the echo, you will be able to see that the raw SQL queries to create those tables will be exactly the same as what we wrote in Listing 5-3.

Listing 5-7. SQLAlchemy metadata–based DDL statements

```
from sqlalchemy import Table, Column,UniqueConstraint, Integer, String,
DateTime, MetaData, ForeignKey

metadata = MetaData()

crawls = Table('crawls', metadata,
    Column('crawl_id', Integer, primary_key=True),
    Column('crawl_url', String, unique=True),
    Column('crawl_desc', String),
    Column('crawl_date', DateTime),
    )

sources = Table('sources', metadata,
    Column('source_id', Integer, primary_key=True),
    Column('source_name', String, unique=True),
    Column('source_url', String, unique=True),
    Column('source_description', String),
    )

webpages = Table('webpages', metadata,
    Column('webpage_id', Integer, primary_key=True),
    Column('crawl_id', None, ForeignKey('crawls.crawl_id')),
    Column('webpage_url', String),
    Column('source_id', None, ForeignKey('sources.source_id')),
    UniqueConstraint('webpage_url', 'crawl_id', name='unique_webpage_
    crawl')
    )

emails = Table('emails', metadata,
    Column('email_id', Integer, primary_key=True),
    Column('email_address', String, unique=True),
    Column('source_id', None, ForeignKey('sources.source_id')),
    )
```

```python
email_webpages = Table('email_webpages', metadata,
    Column('email_webpages_id', Integer, primary_key=True),
    Column('webpage_id', None, ForeignKey('webpages.webpage_id')),
    Column('email_id', None, ForeignKey('emails.email_id')),
    UniqueConstraint('webpage_id', 'email_id', name='unique_webpage_email')
    )

articles = Table('articles', metadata,
    Column('article_id', Integer, primary_key=True),
    Column('published_date', DateTime),
    Column('title', String),
    Column('text', String),
    Column('image_url', String),
    Column('topic', String),
    Column('webpage_id', None, ForeignKey('webpages.webpage_id'))
    )

persons = Table('persons', metadata,
    Column('person_id', Integer, primary_key=True),
    Column('full_name', String, unique=True),
    Column('first_name', String),
    Column('last_name', String),
    Column('source_id', None, ForeignKey('sources.source_id'))
    )

article_authors = Table('article_authors', metadata,
    Column('article_author_id', Integer, primary_key=True),
    Column('webpage_id', None, ForeignKey('webpages.webpage_id')),
    Column('person_id', None, ForeignKey('persons.person_id')),
    UniqueConstraint('webpage_id', 'person_id', name='unique_article_
    authors')
    )
metadata.create_all(engine)
print(metadata.tables.keys())

# Output

dict_keys(['crawls', 'sources', 'webpages', 'emails', 'email_webpages',
'articles', 'persons', 'article_authors'])
```

SQLAlchemy has loads of other advantages apart from using it to make DDL statements for writing portable SQL, and we use it widely at Specrom Analytics. However, we always use it to complement our SQL statements and not as a substitute for the SQL language itself.

In the past few years, I have noticed a disturbing trend of the lack of SQL language knowledge among newly hired data analysts. They seem to rely almost completely on ORM support which abstracts away all the SQL language without really understanding what's happening inside the hood. This makes it very difficult to troubleshoot any ORM-related bugs. I would caution against such an approach and recommend that everyone maintain some fluency in SQL. Its syntax is pretty easy to pick up; the language has been around for over 30 years so it's prevalent everywhere from big data toolchains to traditional databases. If you plan to work in the data analytics field, then I think it is in your best interest to learn the ANSI standard SQL language itself so that you can use it throughout your career irrespective of whether you change your ORM library or software language 10–20 years down the road.

Hence, we will stick to using raw SQL queries as well as its equivalent in SQLAlchemy's expression language but refrain from using their ORM itself. SQLAlchemy makes it really simple to directly send over raw SQL queries using the text() method shown as follows:

```
from sqlalchemy.sql import text

conn = engine.connect()

s = text(
"RAW SQL QUERIES;"
)

conn.execute(s)
```

You should never run unsanitized SQL queries directly to avoid being vulnerable to SQL injection attacks. The comic *xkcd* has done a humorous take on this (https://xkcd.com/327/). If the SQL query in the preceding example is being sent to the database by SQLAlchemy, then it's better to separate out the parameter from the query itself so that the query becomes robust against SQL injection attacks.

For example, in the following query, we want to pick all rows from table_1 in case col_name1 has a value higher than 10. In order to write a safe SQL statement, we are passing the value 10 as a separate parameter (e1) so that our database knows the difference between parameters and SQL statements.

```
from sqlalchemy.sql import text
conn = engine.connect()
s = text(
"SELECT * FROM table_1 WHERE col_name1 > :e1;"
)
conn.execute(s, e1= 10)
```

Data manipulation language (DML) and Data Query Language (DQL)

SQL statements which deal with adding or inserting, updating, and deleting data in a database are known as data manipulation language (DML). Read-only selecting of data is performed by SQL statements known as Data Query Language (DQL).

Most Python developers are pretty well versed in the pandas dataframe so they should check out this document which outlines equivalent DQL commands with the pandas dataframe's operations (https://pandas.pydata.org/docs/getting_started/comparison/comparison_with_sql.html).

One of the most used DML clauses is INSERT INTO which is used to add data in a table. Single quotation marks are required for enclosing character and date/time data types; it's not needed for NULL and numeric type.

```
INSERT INTO table_name
VALUES ('value1', 'value2',...);
```

The preceding statement will insert data into columns in the same order in which the columns were initially created. It's better to explicitly specify the column names where the data is to be inserted.

The general format of a DQL language statement is shown in the following. You use the SELECT clause to pick specific data columns, mentioning the name of tables in the FROM clause with some conditions attached in the WHERE clause. The GROUP BY clause works in the same way as the pandas dataframe's groupby method, and HAVING is analogous to how WHERE is used in a normal SELECT statement. Lastly, ORDER BY simply mentions the sort order of the results. The default behavior is the ascending order (ASC); a descending order can be specified explicitly by the keyword DESC at the end of the ORDER BY clause. In the case of ordering by multiple columns, specify the DESC and ASC keywords next to each column name. We can use numbers in the ORDER BY clause to refer to columns without typing them out explicitly.

```
SELECT col_name1, col_name2
FROM table_name1, table_name2
WHERE (some condition)
GROUP BY col_name1, col_name2
HAVING (some condition)
ORDER BY col_name1, col_name2;
```

A very useful type of DQL is known as "join" statements; they simply merge two tables based on values on foreign keys. The most common type of join is called an equijoin or an inner join where you only get rows which are present in both tables. The pandas merge method will perform equijoin by default. The syntax is shown below; note that T1.col_name1 and T2.col_name2 share a foreign key relationship. The other ways to join it would be the left outer join or right outer join where all values from the left or right table are shown, or you can do a full join where all values from both tables are shown.

```
-- example of equijoin

SELECT  T1.col_name1, T2.col_name2, T2.col_name3
FROM table_1 AS T1, table_2 AS T2
WHERE T1.col_name1 = T2.col_name2 AND T2.col_name3 > 10;

-- this is equivalent to innerjoin syntax

SELECT  T1.col_name1, T2.col_name2, T2.col_name3
FROM  table_1 AS T1 INNER JOIN table_2 AS T2
ON  T1.col_name1 = T2.col_name2
WHERE T2.col_name3 > 10;
```

The typical way of filling up a database is bringing in data from an external source such as a CSV file onto a temporary SQL table with no constraints known as a staging table. Once we have the data we need on the database itself, we will select a subset of data as per our needs using DQL statements and start inserting that into the permanent tables we have created as part of DDL statements in the previous section. Once all the data is transferred to our permanent tables, we can delete the staging table.

Now that we have all the major DQL elements, let us show a sample DML query which moves data over from the staging table to permanent ones.

```
INSERT OR IGNORE INTO table_1 (col_name1, col_name2) "
SELECT T2.col_name1, T3.col_name2
FROM table_2 as T2, table_3 as T3
WHERE some condition;
```

The preceding query will ignore any conflicts on the primary keys or unique constraints and will move on to trying to insert other rows. This can be an expected behavior in many situations.

In a minority of cases, we may want to delete an existing row on conflict and insert a fresh row with possibly data in more columns. This can be achieved with a statement of the type "INSERT or UPDATE"; however, this can mess your foreign key assignments in other tables so you should use this with care.

A more appropriate solution is performing an insert using INSERT OR IGNORE statements and writing a second UPDATE statement which will perform a row update on a scenario when some condition is met as shown in the following. This will keep the primary keys of the rows unaffected and only update the columns mentioned in the SET clause.

```
UPDATE table_name SET col_name1 = (select subquery which shortlists rows
for performing updates);
```

Such an insert followed by an update operation is known as "upsert," and fortunately, lots of RDBMS now support this statement out of the box so that you can run one query insert of an insert statement followed by an update. This is generally done by adding the clause "ON CONFLICT" followed by "SET UPDATE" to the insert clause so that we mention the update criteria directly in one statement itself.

SQLite has added an UPSERT support from versions 3.24.0 and higher; however, most SQLite distributions and existing databases in embedded applications are still running on version 3.21.0 or lower, and it takes some effort to properly upgrade the default SQLite version of your operating system. Hence, in this book, I have refrained from using the UPSERT SQLite statement, but you are encouraged to check it out.

Let us look at the DELETE statement to round off our DML discussion; this allows us to delete specific rows from a given table where certain conditions from a specified column are met. If you want to delete all rows from a table, then the TRUNCATE statement from DDL is a faster way to do it.

```
DELETE FROM table_1
WHERE col_name_id IN (select subquery to pick specific rows);
```

If you already have a metadata object from SQLAlchemy, then the following statement will delete all tables in a database with automatic DROP statements (as seen in the DDL section).

```
metadata.drop_all(engine)
```

This is a very risky way to delete data so be very careful on calling drop_all. A much more calibrated query is deleting by picking specific rows.

Data insertion in SQLite

Let us take a sample file containing email addresses and other additional details such as crawl date and so on and load it to the database as a staging table.

We should convert dates into Python date types as shown in Listing 5-8 so that we don't get any errors due to implicit data type conversions down the road.

Listing 5-8. Pandas dataframe for database insertion

```
import pandas as pd
import numpy as np
from datetime import date

df = pd.read_csv("emails_db_ready.csv")
df["crawl_date"] = pd.to_datetime(df["crawl_date"]).dt.date
df.head()
```

#Output

	crawl_date	crawl_desc	crawl_url	email	email_base_url	webpage_source	webpage_url
0	2020-05-01	us fda warning letters	S3bucketlocation/keyname	Lynn.Bonner@fda.hhs.gov	fda.hhs.gov	fda.gov	https://www.fda.gov/inspections-compliance-enf...
1	2020-05-01	us fda warning letters	S3bucketlocation/keyname	Lynn.Bonner@fda.hhs.gov	fda.hhs.gov	fda.gov	https://www.fda.gov/inspections-compliance-enf...
2	2020-05-01	us fda warning letters	S3bucketlocation/keyname	feb@fda.hhs.gov	fda.hhs.gov	fda.gov	https://www.fda.gov/inspections-compliance-enf...
3	2020-05-01	us fda warning letters	S3bucketlocation/keyname	feb@fda.hhs.gov	fda.hhs.gov	fda.gov	https://www.fda.gov/inspections-compliance-enf...
4	2020-05-01	us fda warning letters	S3bucketlocation/keyname	alan@thepipeshop.co.uk	thepipeshop.co.uk	fda.gov	https://www.fda.gov/inspections-compliance-enf...

Pandas has a very useful method called to_sql which allows us to pass a SQLAlchemy engine. We can also explicitly pass in data types with the dtypes parameter. If your pandas version is 0.24 or newer, there is a parameter called methods, and selecting "multi" will allow you to pass in multiple values per commit instead of one per insert clause like usual.

I recommend that you should always check the data types of the newly created staging table as shown in Listing 5-9 and make sure it conforms to your needs.

Listing 5-9. Creating a staging table

```
from sqlalchemy import MetaData, DateTime

metadata = MetaData()
metadata.reflect(engine)
df.to_sql('staging',con = engine, dtype={"crawl_date":DateTime})

metadata.reflect(engine)
metadata.tables["staging"]
```

Output

```
Table('staging', MetaData(bind=None), Column('index', BIGINT(),
table=<staging>), Column('crawl_date', DATETIME(), table=<staging>),
Column('crawl_desc', TEXT(), table=<staging>), Column('crawl_url',
TEXT(), table=<staging>), Column('email', TEXT(), table=<staging>),
Column('email_base_url', TEXT(), table=<staging>), Column('webpage_source',
TEXT(), table=<staging>), Column('webpage_url', TEXT(), table=<staging>),
schema=None)
```

We can now start using a DML statement called insert into to move data from staging to a couple of main tables as shown in Listing 5-10. SQLAlchemy has a text() function which lets us pass raw SQL statements to the engine.

We will use the ignore clause to handle cases when a duplicate entry already exists in our main tables. The SELECT statement is part of DQL, but here all it's telling our database is that it should copy specific columns from staging and insert it into the columns in parentheses in crawl and sources tables.

One interesting SQL statement used here is UNION which simply concatenates results from two different SELECT statements. Here, it's used so that we can populate a single column (sources_url) from data present in two columns in staging.

Listing 5-10. Inserts from the staging table

```
from sqlalchemy.sql import text
conn = engine.connect()

insert_into_crawls_table = text(
"INSERT OR IGNORE INTO crawls (crawl_date, crawl_url, crawl_desc) "
"SELECT crawl_date, crawl_url, crawl_desc FROM staging;"

)
conn.execute(insert_into_crawls_table)

insert_into_sources_table = text(
"INSERT OR IGNORE INTO sources (source_url) "
"SELECT webpage_source FROM staging UNION SELECT email_base_url FROM
staging;"

)

conn.execute(insert_into_sources_table)
```

The next insert statement into the webpages table gets a bit complicated as shown in Listing 5-11. We want to insert webpage urls from the staging table into the webpages table along with filling out foreign key reference columns such as crawl_id and source_id.

Listing 5-11. Inserts from the staging table

```
insert_into_webpages_table = text(
"INSERT OR IGNORE INTO webpages (crawl_id, webpage_url, source_id) "
"SELECT crawls.crawl_id, staging.webpage_url, sources.source_id "
"FROM crawls, staging, sources "
"WHERE staging.crawl_url = crawls.crawl_url "
"AND staging.webpage_source = sources.source_url;"
)

conn.execute(insert_into_webpages_table)
```

We can wrap up all our inserts by filling up emails and email_webpages tables shown in Listing 5-12.

Listing 5-12. Inserts from the staging table

```
insert_into_emails_table = text(
"INSERT OR IGNORE INTO emails (source_id, email_address) "
"SELECT sources.source_id, staging.email "
"FROM sources, staging "
"WHERE staging.email_base_url = sources.source_url;"

)
conn.execute(insert_into_emails_table)

insert_into_email_webpages_table = text(
"INSERT OR IGNORE INTO email_webpages (webpage_id, email_id) "
"SELECT webpages.webpage_id, emails.email_id "
"FROM webpages, emails,staging "
"WHERE staging.webpage_url = webpages.webpage_url "
"AND staging.email = emails.email_address;"

)
conn.execute(insert_into_email_webpages_table)

drop_staging_table = text(
    "DROP TABLE staging;"
    )
conn.execute(drop_staging_table)
```

Now that we have the emails-related tables filled up, let us query the database in Listing 5-13 to show all the email addresses and the web pages where they were found by passing the domain address as a parameter. If you were building a Hunter.io type clone, then all you need now is expose this query as an API and package the following response into a JSON.

Listing 5-13. DQL statements for fetching email addresses from the database

```
from sqlalchemy.sql import text

select_emails = text(
"SELECT emails.email_address, webpages.webpage_url "
"FROM emails, webpages, email_webpages "
"WHERE emails.source_id = (select sources.source_id from sources where
source_url = :e1) AND "
"emails.email_id = email_webpages.email_id AND "
"email_webpages.webpage_id = webpages.webpage_id "
"ORDER BY emails.email_address;"
)

query = conn.execute(select_emails, e1 = 'fda.hhs.gov')
result_list = query.fetchall()
result_list_keys = query.keys() #https://stackoverflow.com/
questions/6455560/how-to-get-column-names-from-sqlalchemy-result-
declarative-syntax
result_list

#Output

[('Araceli.rey@fda.hhs.gov', 'https://www.fda.gov/inspections-compliance-
enforcement-and-criminal-investigations/warning-letters/apotex-research-
private-limited-437669-01302015'),
 ('CTPCompliance@fda.hhs.gov', 'https://www.fda.gov/inspections-compliance-
 enforcement-and-criminal-investigations/warning-letters/pipe-shop-
 edinburgh-01072015'),
 ('CTPCompliance@fda.hhs.gov', 'https://www.fda.gov/inspections-
 compliance-enforcement-and-criminal-investigations/warning-letters/usa-
 cigscom-01162015'),
```

```
('CTPCompliance@fda.hhs.gov', 'https://www.fda.gov/inspections-
 compliance-enforcement-and-criminal-investigations/warning-letters/
 wwwcigs24x7com-02052015'),
('Lillian.Aveta@fda.hhs.gov', 'https://www.fda.gov/inspections-compliance-
 enforcement-and-criminal-investigations/warning-letters/nysw-beverage-
 brands-438025-01072015'),
('Lynn.Bonner@fda.hhs.gov', 'https://www.fda.gov/inspections-compliance-
 enforcement-and-criminal-investigations/warning-letters/nestle-purina-
 petcare-01022015'),
('ReportabilityReviewTeam@fda.hhs.gov', 'https://www.fda.gov/inspections-
 compliance-enforcement-and-criminal-investigations/warning-letters/
 praxair-inc-01072015'),
('ReportabilityReviewTeam@fda.hhs.gov', 'https://www.fda.gov/inspections-
 compliance-enforcement-and-criminal-investigations/warning-letters/
 conkin-surgical-instruments-ltd-01122015'),
('drugshortages@fda.hhs.gov', 'https://www.fda.gov/inspections-compliance-
 enforcement-and-criminal-investigations/warning-letters/apotex-research-
 private-limited-437669-01302015'),
('drugshortages@fda.hhs.gov', 'https://www.fda.gov/inspections-compliance-
 enforcement-and-criminal-investigations/warning-letters/micro-labs-
 limited-01092015'),
('feb@fda.hhs.gov', 'https://www.fda.gov/inspections-compliance-
 enforcement-and-criminal-investigations/warning-letters/sagami-rubber-
 industries-co-ltd-11242015'),
('feb@fda.hhs.gov', 'https://www.fda.gov/inspections-compliance-
 enforcement-and-criminal-investigations/warning-letters/prodimed-
 sas-481475-11022015'),
('feb@fda.hhs.gov', 'https://www.fda.gov/inspections-compliance-
 enforcement-and-criminal-investigations/warning-letters/oculo-plastik-
 inc-12222015'),
('matthew.dionne@fda.hhs.gov', 'https://www.fda.gov/inspections-
 compliance-enforcement-and-criminal-investigations/warning-letters/sunny-
 boys-dairy-01132015')]
```

We can also query for all the email addresses in our database, and once we have them in memory, we can use pandas value counts to check for common mailbox names in an email address (the part before @ in an email address). In our experience, common mailbox names are info, contact, hello, and so on, especially among email databases composed of crawled web pages. In this case, though, since we only crawled through one data source (the US FDA warning letters), abuse seems to be the common mailbox name.

Listing 5-14. Analyzing email addresses

```
import pandas as pd
import numpy as np

fetch_all_emails = text(
    "select emails.email_address, emails.email_id from emails;"
    )
query = conn.execute(fetch_all_emails)
result_list = query.fetchall()
result_list_keys = query.keys()

df = pd.DataFrame(result_list, columns = result_list_keys)
df["mailbox_name"] = df["email_address"].apply(lambda x: x.split('@')[0])
df.mailbox_name.value_counts().head()

#Output

abuse                    5
Yvette.Johnson           1
ReportabilityReviewTeam  1
Lynn.Bonner              1
drugshortages            1
Name: mailbox_name, dtype: int64
```

Inserting other tables

We still need to fill the table with full text from a web page. For this purpose, we will use another CSV file which contains extracted dates, full text, titles, author names (if any), and so on. Note that we have used the UPDATE DQL statement for the sources table to perform an upsert type operation as shown in Listing 5-15.

Listing 5-15. Inserting data into other tables

```python
import pandas as pd
import numpy as np
from datetime import date

df = pd.read_csv("news_data.csv")
df["crawl_date"] = pd.to_datetime(df["crawl_date"]).dt.date
df['date'].replace('None', np.nan, inplace=True)
df["date"] = df["date"].fillna(df["crawl_date"])
df["date"] = pd.to_datetime(df["date"]).dt.date

from sqlalchemy import MetaData, DateTime

metadata = MetaData()
metadata.reflect(engine)
df.to_sql('staging',con = engine, dtype={"crawl_date":DateTime,
"date":DateTime})

metadata.reflect(engine)
metadata.tables["staging"]

from sqlalchemy.sql import text

insert_into_crawls_table = text(
"INSERT OR IGNORE INTO crawls (crawl_date, crawl_url, crawl_desc) "
"SELECT crawl_date, crawl_url, crawl_desc FROM staging;"

)
conn.execute(insert_into_crawls_table)

insert_into_sources_table = text(
"INSERT OR IGNORE INTO sources (source_url, source_name) "
"SELECT webpage_source, site_name FROM staging;"
)
conn.execute(insert_into_sources_table)

update_sources = text(
"UPDATE sources SET source_name = (SELECT staging.site_name FROM staging
WHERE source_url=staging.webpage_source);"
)
```

```
conn.execute(update_sources)

insert_into_webpages_table = text(
"INSERT OR IGNORE INTO webpages (crawl_id, webpage_url, source_id) "
"SELECT crawls.crawl_id, staging.url, sources.source_id "
"FROM crawls, staging, sources "
"WHERE staging.crawl_url = crawls.crawl_url "
"AND staging.webpage_source = sources.source_url;"
)

conn.execute(insert_into_webpages_table)

insert_into_articles_table = text(
"INSERT OR IGNORE INTO articles (published_date,title, text, image_
url,topic, webpage_id) "
"SELECT staging.date, staging.title, staging.full_text, staging.img_url,
staging.topic, webpages.webpage_id "
"FROM webpages, staging "
"WHERE staging.url = webpages.webpage_url;"

)
conn.execute(insert_into_articles_table)

insert_into_persons_table = text(
"INSERT OR IGNORE INTO persons (full_name) "
"SELECT staging.author "
"FROM staging "
"WHERE staging.author IS NOT NULL;"

)
conn.execute(insert_into_persons_table)

insert_into_article_authors_table = text(
"INSERT OR IGNORE INTO article_authors (webpage_id, person_id) "
"SELECT webpages.webpage_id, persons.person_id "
"FROM persons, staging, webpages "
"WHERE staging.author IS NOT NULL "
```

```
"AND staging.author = persons.full_name "
"AND staging.url = webpages.webpage_url;"

)
conn.execute(insert_into_article_authors_table)

drop_staging_table = text(
"DROP TABLE staging;"
)
conn.execute(drop_staging_table)
```

You can perform basic full text searching over multiple columns using DQL statements which concatenate multiple columns and use the LIKE operator to find token matches as shown in Listing 5-16.

Listing 5-16. Full text searching using the LIKE operator

```
from sqlalchemy.sql import text
import pandas as pd
import numpy as np

s = text(
"SELECT articles.article_id, webpages.webpage_url, "
"articles.title || ' ' || articles.text || ' ' || articles.topic AS
fulltext "
"FROM articles INNER JOIN webpages ON articles.webpage_id = webpages.
webpage_id "
"WHERE fulltext LIKE '%trump%' "
"AND fulltext LIKE '%losing%';"
)
conn = engine.connect()
query = conn.execute(s)
result_list = query.fetchall()
result_list_keys = query.keys()

df = pd.DataFrame(result_list, columns = result_list_keys)
df.head()
```

Output

article_id	webpage_url	fulltext
0 7	https://www.iolaregister.com/ opinion/americans...	Americans losing confidence in democracy - The...

This suffers from a host of problems. Firstly, it will not find matches for stemmed versions of the tokens such as lose instead of losing; the query itself seems pretty verbose, and we don't get a relevance score to the search query which is expected out of a full text searching database such as Lucene or Elasticsearch. Hence, let us look at a specific SQLite module which helps us build a search index for a better search performance.

Full text searching in SQLite

Full text search support in SQLite is provided by the module called FTS5, and it is more powerful than you would expect for a lightweight library; it even supports porter stemming–based tokenization.

FTS5 has been available from SQLite version 3.9.0, so almost all SQLite installations should have it already. If you are having an older version of SQLite for whatever reason, then I highly recommend that you look into upgrading it.

We created a virtual table in Listing 5-17 with the FTS5 plugin which uses porter tokenizer; it fills contents based on concatenated contents from the articles table.

Listing 5-17. Creating virtual table using FTS5

```
create_fts_table = text("CREATE VIRTUAL TABLE news_index USING fts5
(fulltext, tokenize=porter);")
conn.execute(create_fts_table)
```

We can insert data into this index as shown in Listing 5-18.

Listing 5-18. Inserting data into the full text search table

```
s = text(
'''INSERT OR IGNORE INTO news_index (
    rowid,
    fulltext
  )
  SELECT
    articles.article_id,
    articles.title || ' ' || articles.text || ' ' || articles.topic AS
fulltext
FROM articles;'''
)
conn.execute(s)
```

A much more robust solution is using triggers to automatically fill the index whenever data in our articles table is inserted, deleted, or updated as shown in Listing 5-19.

Listing 5-19. Trigger on insert and updates in the articles table

```
insert_trigger = text(
'''CREATE TRIGGER fts_articles_insert AFTER INSERT ON articles BEGIN
  INSERT INTO news_index (
    rowid,
    fulltext
  )
  SELECT
    articles.article_id,
    articles.title || ' ' || articles.text || ' ' || articles.topic AS
fulltext
FROM articles;
END;'''
)
conn.execute(insert_trigger)

# trigger on update on articles table
```

```
update_trigger = text(
'''CREATE TRIGGER fts_articles_update UPDATE OF news_index ON articles
BEGIN
  UPDATE news_index SET fulltext = new.fulltext WHERE rowid = old.id;
END;''')
conn.execute(update_trigger)

# trigger on delete on articles table

delete_trigger = text(
'''CREATE TRIGGER fts_articles_delete AFTER DELETE ON articles BEGIN
    DELETE FROM news_index WHERE rowid = old.id;
END;'''
)
conn.execute(delete_trigger)
```

Lastly, let's test out the index and query for the rank score in Listing 5-20; note that we will use a base word inflection of "losing" here ("lose") and see if our index produces the same result as before.

Listing 5-20. Full text searching using FTS5

```
s = text(
    "SELECT rowid, rank, fulltext "
    "FROM news_index "
    "WHERE news_index MATCH 'fulltext:Trump AND lose' ORDER BY rank;"
)
query = conn.execute(s)
result_list = query.fetchall()
result_list_keys = query.keys()

df = pd.DataFrame(result_list, columns = result_list_keys)
df.head()

# Output
```

	rowid	rank	fulltext
0	7	-3.758945	Americans losing confidence in democracy - The...

Search indexes such as these are generally used behind the scenes so that you get the same result as Listing 5-21 but with stemming and other preprocessing done already.

Listing 5-21. Using full text search index on the back end

```
s = text(
'''SELECT article_id, webpage_url, articles."text"
FROM webpages INNER JOIN articles
ON articles.webpage_id = webpages.webpage_id
WHERE article_id = (SELECT rowid
FROM news_index
WHERE news_index MATCH 'fulltext:Trump AND lose');'''
)
query = conn.execute(s)
result_list = query.fetchall()
result_list_keys = query.keys()

df = pd.DataFrame(result_list, columns = result_list_keys)
df.head()

#Output
```

	article_id	webpage_url	text
0	7	https://www.iolaregister.com/ opinion/americans...	Americans losing confidence in democracy Peop...

One downside is that SQLite FTS doesn't work out of the box with any serverless back ends such as AWS Lambda or even Algorithmia; hence, if you plan to run SQLite on serverless computing platforms, then you'll have to get FTS5 on AWS Lambda using advanced tricks such as creating it from a docker image or switch to a server-based back end for your REST APIs.

Another way would be to simply rely on the LIKE function from ANSI SQL such as shown in Listing 5-21, but that doesn't have quite the same functionality as discussed in this section.

Data insertion in PostgreSQL

PostgreSQL supports an even faster method called COPY which can be used to transfer data from a CSV file to the staging table apart from using the pandas dataframe's to_csv function for SQLite.

If you are using a pandas version 0.24 or higher, then you can pass a callable to the methods parameter which makes use of the PostgreSQL copy clause (more info here: `https://pandas.pydata.org/pandas-docs/stable/user_guide/io.html#io-sql-method`). However, if you aren't running that version yet, then we can achieve the same result by the query shown in Listing 5-22.

Listing 5-22. Data insertion into the staging table

```
import pandas as pd
import numpy as np
from datetime import date
import io

df = pd.read_csv("emails_db_ready.csv")
df["crawl_date"] = pd.to_datetime(df["crawl_date"]).dt.date
df.head()

df.head(0).to_sql('staging', engine, if_exists='replace',index=False)
raw_conn = engine.raw_connection()
cur = raw_conn.cursor()
output = io.StringIO()
df.to_csv(output, sep='\t', header=False, index=False)
output.seek(0)

cur.copy_from(output, 'staging', null="") # null values become ''
raw_conn.commit()
```

If you are using AWS RDS–based PostgreSQL, then your input/output operations per second (IOPS) are throttled based on the size of the disk and the plan you are currently subscribed.

Another PostgreSQL-related common issue is making sure that we do not try to insert text with a large number of characters within a column with a unique constraint. In our schema, this might happen with the unique_webpage_crawl constraint which requires that the combination of webpage_url and crawl_id columns is unique. The error you might get will be something like "index row size XXX exceeds maximum 2712 for index 'unique_webpage_crawl.'"

In such a case, you can either truncate the webpage urls to stay below the limit of 2712 characters, delete any rows with longer webpage_urls than this number, or alternately create an MD5 hash of the webpage_url using the MD5() function in PostgreSQL and place a unique constraint on that instead of the URL itself; there are few other workarounds and you can check them out here (https://dba.stackexchange.com/questions/162820/values-larger-than-1-3-of-a-buffer-page-cannot-be-indexed).

We have decided not to do major PostgreSQL-related changes to the database schema so that we keep it completely equivalent to the SQLite-based schema, but I wanted to highlight these common PostgreSQL-related issues so that you are aware of them.

Once data is in the staging table, there is minimal difference in inserting it in permanent tables here vs. SQLite as shown in Listing 5-23. We have used the upsert format here since PostgreSQL versions higher than 9.5 have been supporting it out of the box.

We also had to explicitly modify the staging table to ensure that the crawl date belongs to the correct PostgreSQL data type (timestamp without time zone).

Listing 5-23. Data insertion into other tables

```
s = text(
"ALTER TABLE staging "
"ALTER COLUMN crawl_date TYPE TIMESTAMP WITHOUT TIME ZONE USING crawl_
date::timestamp without time zone;"
)

conn.execute(s)
```

```
insert_into_crawls_table = text(
"INSERT INTO crawls (crawl_date, crawl_url, crawl_desc) "
"SELECT crawl_date, crawl_url, crawl_desc FROM staging "
"ON CONFLICT DO NOTHING;"

)
conn.execute(insert_into_crawls_table)

insert_into_sources_table = text(
"INSERT INTO sources (source_url) "
"SELECT webpage_source FROM staging "
"ON CONFLICT DO NOTHING;"

)
conn.execute(insert_into_sources_table)

insert_into_webpages_table = text(
"INSERT INTO webpages (crawl_id, webpage_url, source_id) "
"SELECT crawls.crawl_id, staging.webpage_url, sources.source_id "
"FROM crawls, staging, sources "
"WHERE staging.crawl_url = crawls.crawl_url "
"AND staging.webpage_source = sources.source_url "
"ON CONFLICT DO NOTHING;"
)

conn.execute(insert_into_webpages_table)

insert_into_emails_table = text(
"INSERT INTO emails (source_id, email_address) "
"SELECT sources.source_id, staging.email "
"FROM sources, staging "
"WHERE staging.email_base_url = sources.source_url "
"ON CONFLICT DO NOTHING;"

)
conn.execute(insert_into_emails_table)

insert_into_email_webpages_table = text(
"INSERT INTO email_webpages (webpage_id, email_id) "
```

```
"SELECT webpages.webpage_id, emails.email_id "
"FROM webpages, emails,staging "
"WHERE staging.webpage_url = webpages.webpage_url "
"AND staging.email = emails.email_address "
"ON CONFLICT DO NOTHING;"

)
conn.execute(insert_into_email_webpages_table)
```

You can test out the PostgreSQL database with the same queries as Listings 5-13 and 5-14 from the SQLite section since DQL queries without any database-specific functions are pretty portable. We have also kept DML queries for inserting data from news_data.csv into PostgreSQL in the source code folder, but it's not shown here.

Full text searching in PostgreSQL

Full text searching in PostgreSQL is extremely powerful, and it supports all major languages such as German, French, Italian, and Portuguese apart from English.

You can view the default text search language by typing the SQL query shown as follows:

```
SHOW default_text_search_config;
```

If it's not the language you want, or if it shows "pg_catalog.simple", then you can set the default language for the session as shown in the following. Alternately, you can alter your database role (for user postgres in the following example) and make pg_catalog. english as default.

```
set default_text_search_config = 'pg_catalog.english';
ALTER ROLE postgres IN DATABASE masterdatabase
SET default_text_search_config TO 'pg_catalog.english';
```

We have discussed in Chapter 4 how natural languages should be converted into numerical vectors and preprocessed by performing stemming or lemmatizations and so on. PostgreSQL lets you query for the word vectors directly using "to_tsvector" shown in Listing 5-24, and you can see clearly that it has stemmed lots of words such as volunteer, confidence, and so on.

Listing 5-24. Testing the to_tsvector function

```
s = text(
'''SELECT to_tsvector('english','Americans losing confidence in democracy
People are giving up on traditional,
political actions like voting, ngs, and volunteering. They see a government
on autopilot,
with little regard to what voters or even politicians want.');''')
conn.execute(s).fetchall()
```

```
# Output
```

```
[("'action':13 'american':1 'autopilot':24 'confid':3 'democraci':5
'even':32 'give':8 'govern':22 'like':14 'littl':26 'lose':2 'ngs':16
'peopl':6 'polit':12 'politician':33 'regard':27 'see':20 'tradit':11
'volunt':18 'vote':15 'voter':30 'want':34",)]
```

We use ts_vector to convert our text into vectors which can be used to concatenate on multiple columns. This can be used to search using the special operator @@. Lastly, it can be ranked based on the relevance score which is generated by ts_query. We combined this in Listing 5-25, and it returned the same results as Listing 5-21.

Listing 5-25. Full text searching in PostgreSQL

```
s = text('''SELECT pid, p_title, p_text, ts_rank(full_search.sample_
document, to_tsquery('english', 'lose & Trump')) as relevancy
FROM (SELECT articles.article_id as pid,
              articles.title as p_title,
              articles.text as p_text,
              to_tsvector('english', articles.title) ||
              to_tsvector('english', articles.text)
              as sample_document
      FROM articles
      GROUP BY articles.article_id) full_search
      WHERE full_search.sample_document @@ to_tsquery('english','lose &
      Trump')
      ORDER BY relevancy DESC;''')
conn.execute(s).fetchall()
```

PostgreSQL full text searching includes quite a few advanced functionalities such as the ability to handle misspelling and assigning differential weightages to individual concatenated columns in the preceding sample_document for computing relevancy scores so that you can place a higher weightage to words which appear in title vs. those which appear in full text.

I don't expect you to grasp all the details about implementing full text search in PostgreSQL since that will take another chapter worth of information, but I hope that through this section you have gotten a small sample of what PostgreSQL can do in this area.

Why do NoSQL databases exist?

PostgreSQL is akin to a Swiss Army knife since apart from a normal relational data storage, it has decent support for storing JSON, full text search, graph database support through AgensGraph, and spatial and geographical object support through an extension called PostGIS.

If you had to only choose one database to take with you on a trip to a desert island, then it should probably be PostgreSQL (although what you will do with a database there is a completely different question!). This may seem like a silly analogy, but that's what I had to face when trying to pick just one full-fledged RDBMS system for inclusion in this chapter.

However, in the real world, we seldom have to make a decision on picking only one tool of a particular kind based on such a false dichotomy, and if you face any scalability, latency, or other issues mentioned in the following, then you should look into NoSQL databases.

A relational data model is sometimes found to be an inefficient way to store certain kinds of data which can be modeled and queried better for intended use cases using NoSQL databases like a graph database such as Neo4j or a full text searching database such as Elasticsearch or Amazon CloudSearch.

As your database size and number of operations go up, you will have to scale up the database. You can either scale up horizontally, where you simply add more servers (known as nodes) and create horizontal partitions of data known as shards, or you can scale up vertically by adding more CPU cores and RAM to one server. Generally speaking, vertical scaling costs a lot more than horizontal scaling, and it also makes systems less fault tolerant, meaning that if one server goes down, your entire database goes down. Traditionally, relational databases such as PostgreSQL were only able to scale vertically, and this is where a lot of NoSQL databases have a clear advantage.

Relational databases are pretty inflexible when it comes to database schema changes, and that is definitely an impediment when it comes to storing data from web scraping where we may need to alter the database to match the structure of the website scraped. You can mitigate this to some extent by storing documents as JSON in postgres, but a better idea might be to switch to a document database such as Amazon DocumentDB or MongoDB especially since it also offers ACID compliance and a host of other features (`www.mongodb.com/compare/mongodb-postgresql`).

Another area where NoSQL databases such as Cassandra and AWS DynamoDB really shine is the low latency for read-only data queries. This is a direct byproduct of their database design which doesn't include support or strongly discourages DQL type joins, making the tables highly denormalized which in turn improves speed.

Summary

We introduced relational databases such as SQLite and PostgreSQL and the SQL language in this chapter. We used them to create a database schema appropriate for storing individual assets from web scraping batch jobs.

Lastly, we demonstrated how to set up full text searching and showed how to write SQL queries to fetch records from the database.

This chapter wraps up all the basic elements of web scraping, and now we can put into use what we have learned into extracting structured information from terabyte scale–sized web corpuses from Common Crawl in the next two chapters.

We will also use Amazon Athena in the next chapter to query for data on TB scale directly in S3 using the SQL language.

CHAPTER 6

Introduction to Common Crawl Datasets

In this chapter, we'll talk about an open source dataset called common crawl which is available on AWS's registry of open data (`https://registry.opendata.aws/`).

AWS hosts a large variety of open datasets on its servers which are freely available to all users. These datasets are uploaded and maintained by third parties, and AWS simply waives off the monthly charges and/or server fees to support these organizations.

The Common Crawl Foundation (`https://commoncrawl.org/`) is a 501(c)(3) nonprofit involved in providing open access web crawl data going back to over eight years. They perform monthly web crawls which cover over 25 billion pages for each month. This petabyte-scale data corpus is used widely by industry giants including Google (`https://arxiv.org/abs/1804.03235`) and Facebook (`https://arxiv.org/abs/1802.06893`) and academia to benchmark their algorithms, training machine learning models and word vectors, security research, and other natural language processing tasks.

Common crawl has completely open sourced its codebase (`https://github.com/commoncrawl`), and it's mainly a Java-based technology stack using Apache Nutch–based distributed crawler along with Hadoop and Apache Spark for further processing of data.

There is about 80–85% coverage of domain addresses in this dataset in comparison with Alexa's top one million domains or Majestic's top one million domains which is pretty good considering that common crawl bot respects the robot.txt and does not crawl any website such as LinkedIn which blocks all web crawlers except from a few whitelisted user agents. You can view other interesting statistics about the project at their stats page (`https://commoncrawl.github.io/cc-crawl-statistics/plots/tld/comparison.html`).

© Jay M. Patel 2020
J. M. Patel, *Getting Structured Data from the Internet*, https://doi.org/10.1007/978-1-4842-6576-5_6

When we take the common crawl data cumulatively, across monthly crawls since 2008, it represents one of the largest publicly accessible web crawl data corpuses on a petabyte scale, and this is one major reason why it's been used so widely in academia and the industry.

The common crawl project also preprocesses their raw web crawl files and makes extracted text and web page metadata available separately.

We introduce common crawl datasets in this chapter and work with them locally to create a similar web page database and a backlinks database such as what we saw in Chapter 1 with Ahrefs and Alexa.

In Chapter 7, we will cover processing a common crawl dataset on a big data production scale using an AWS-based distributed computing framework based on EC2, S3, SQS, and Athena.

WARC file format

The raw web crawls are stored in web archive (WARC) file format which is optimized for efficient storage and querying for large-sized web corpuses. You can check out the official ISO standard (`www.iso.org/standard/68004.html`) for the WARC file format for more information.

Let's abstract away all the intricacies of the file format–related issues and open them using a Python package called warc. It's available on `https://github.com/jaypat87/warc`; just clone it to your local directory, unzip and enter the cloned folder, and enter the command on bash or the command line "pip install ." to install it on your local computer.

Note that the older versions of the warc package available on pip will not work with current Python versions, and hence it's best to install the package available on GitHub.

Common crawl announces the monthly crawl details as a blog announcement (`https://commoncrawl.org/connect/blog/`); we can go to the March 2020 crawl announcement (`https://commoncrawl.org/2020/04/march-april-2020-crawl-archive-now-available/`) and get the WARC file paths shown in Listing 6-1 for March 2020 crawls. We are only printing the first few hundred bytes, but it contains file paths to 60,000 warc files with a total size of about 62 TB (compressed), and it's available on the S3 bucket in the US-East-1 region.

Listing 6-1. Fetching file paths for warc files

```
import requests
import gzip
url = 'https://commoncrawl.s3.amazonaws.com/crawl-data/CC-MAIN-2020-16/
warc.paths.gz'
from io import BytesIO

r = requests.get(url)
compressed_file = BytesIO(r.content)
f = gzip.GzipFile(fileobj=compressed_file)
print(f.read(326).decode("utf-8"))

# Output
crawl-data/CC-MAIN-2020-16/segments/1585370490497.6/warc/CC-MAIN
-20200328074047-20200328104047-00000.warc.gz
crawl-data/CC-MAIN-2020-16/segments/1585370490497.6/warc/CC-MAIN
-20200328074047-20200328104047-00001.warc.gz
crawl-data/CC-MAIN-2020-16/segments/1585370490497.6/warc/CC-MAIN
-20200328074047-20200328104047-00002.warc.gz
```

Let us download one of the warc files from the S3 bucket in Listing 6-2 and explore it using the warc package. This warc file is about 1.2 GB in size and contains raw HTML page and header information for about 54,000 web pages. These web pages are fetched in pseudo-random order, so web pages belonging to a particular web domain may be split across many hundred or thousand warc files.

Listing 6-2. Downloading a warc file

```
warc_path =  'crawl-data/CC-MAIN-2020-16/segments/1585370490497.6/warc/CC-
MAIN-20200328074047-20200328104047-00455.warc.gz'
file_name = 'YOUR_LOCAL_FILEPATH.warc.gz'

import boto3

from botocore.handlers import import disable_signing
resource = boto3.resource('s3')
resource.meta.client.meta.events.register('choose-signer.s3.*', disable_
signing)
```

```python
bucket = resource.Bucket('commoncrawl')

resource.meta.client.download_file('commoncrawl', warc_path, file_name)

from time import time

import warc

def process_warc(file_name, limit=10000):
    warc_file = warc.open(file_name, 'rb')
    t0 = time()
    n_documents = 0

    url_list = []
    header_list = []
    html_content = []

    for i, record in enumerate(warc_file):

        if n_documents >= limit:

            break

        url = record.url
        payload = record.payload.read()

        try:
            header, html = payload.split(b'\r\n\r\n', maxsplit=1)
            html = html.strip()
        except:

            continue

        if url is None or payload is None or html == b'':

            continue

        else:
            try:
```

```
                html_content.append(html)
                header_list.append(header)
                url_list.append(url)
            except Exception as e:
                #print(e)
                continue

        n_documents += 1

    warc_file.close()
    print('Parsing took %s seconds and went through %s documents' %
(time() - t0, n_documents))
    return header_list, html_content, url_list

file_name = 'YOUR_LOCAL_FILEPATH.warc.gz'
header_list, html_content, url_list = process_warc(file_name, limit =
1000000)
```

Output

Parsing took 45.039262771606445 seconds and went through 54262 documents

Let's explore the headers and HTML from one of the web pages in this warc file as shown in Listing 6-3.

Listing 6-3. Exploring the warc record for a web page

```
print(url_list[867])
print('*'*10)
print(header_list[867])
print('*'*10)
print(html_content[867])
# Output
```

http://archive.griffith.ox.ac.uk/index.php/informationobject/browse?view=ca
rd&languages=en&creators=393&mediatypes=136&sort=referenceCode&sf_culture=e
n&levels=223&topLod=0&limit=30&sortDir=asc
****

```
HTTP/1.1 200 OK
Server: nginx/1.14.0 (Ubuntu)
Date: Sat, 28 Mar 2020 10:01:19 GMT
Content-Type: text/html; charset=utf-8
X-Crawler-Transfer-Encoding: chunked
Connection: keep-alive
Set-Cookie: symfony=jjecpro8lfekf6nmo9hj7qc5eb; path=/; HttpOnly
Expires: Thu, 19 Nov 1981 08:52:00 GMT
Cache-Control: no-store, no-cache, must-revalidate
Pragma: no-cache
X-Ua-Compatible: IE=edge,chrome=1
X-Crawler-Content-Encoding: gzip
Content-Length: 31469
**********
<!DOCTYPE html>
<html lang="en" dir="ltr">
  <head>
    <meta http-equiv="Content-Type" content="text/html; charset=utf-8" />
<meta http-equiv="X-Ua-Compatible" content="IE=edge,chrome=1" />
    <meta name="title" content="Griffith Institute Archive" />

. . .

(html truncated)
```

Common crawl index

Each web page record in a WARC file is compressed and stored in such a way that we can directly fetch individual records using the file offsets which can be queried as part of the common crawl URL index. These index files are about 200 GB and also available in a highly efficient Apache Parquet columnar file format so that you can run the batch queries on Athena which we will see in Chapter 7.

The common crawl index is exposed via a CDX server API (`https://github.com/webrecorder/pywb/wiki/CDX-Server-API`) so that we can easily fetch individual HTML records for a given URL and also check if a particular domain address has been indexed or not.

Just go to the `http://index.commoncrawl.org/` page, and you will see a table with search page and API endpoint information as shown in Figure 6-1. Let us click / CC-MAIN-2020-16 since March 2020 crawls are what we have used in the rest of the examples.

Search Page	Crawl	API endpoint	Index File List on `s3://commoncrawl/`
/CC-MAIN-2020-29	July 2020 Index	/CC-MAIN-2020-29-index	CC-MAIN-2020-29/cc-index.paths.gz
/CC-MAIN-2020-24	May 2020 Index	/CC-MAIN-2020-24-index	CC-MAIN-2020-24/cc-index.paths.gz
/CC-MAIN-2020-16	March 2020 Index	/CC-MAIN-2020-16-index	CC-MAIN-2020-16/cc-index.paths.gz
/CC-MAIN-2020-10	February 2020 Index	/CC-MAIN-2020-10-index	CC-MAIN-2020-10/cc-index.paths.gz
/CC-MAIN-2020-05	January 2020 Index	/CC-MAIN-2020-05-index	CC-MAIN-2020-05/cc-index.paths.gz
/CC-MAIN-2019-51	December 2019 Index	/CC-MAIN-2019-51-index	CC-MAIN-2019-51/cc-index.paths.gz
/CC-MAIN-2019-47	November 2019 Index	/CC-MAIN-2019-47-index	CC-MAIN-2019-47/cc-index.paths.gz

Figure 6-1. *Common crawl index page*

Simply enter the URL of interest in the search bar shown in Figure 6-2. We can fetch all the captures for a particular domain by specifying wildcards (*). For example, if we want to see all the captures for apress.com, simply search for apress.com/*. Similarly, if we want to check for all the subdomains, we can search like this: *.apress.com.

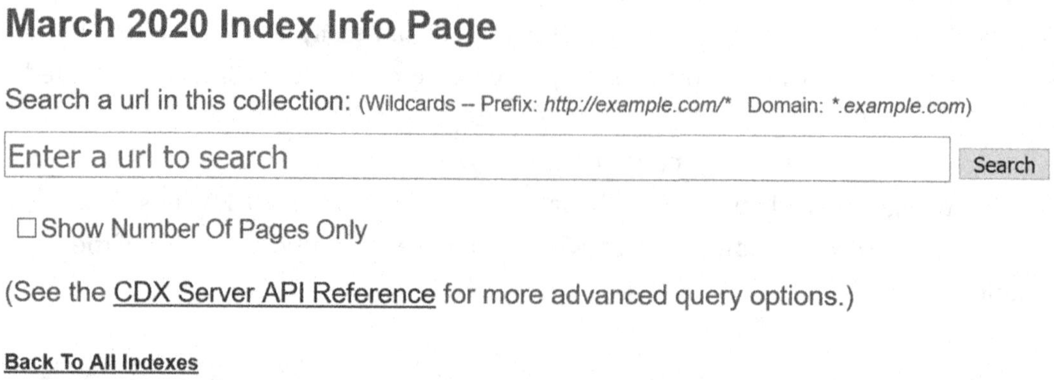

March 2020 Index Info Page

Search a url in this collection: (Wildcards – Prefix: *http://example.com/** Domain: **.example.com*)

Enter a url to search	Search

☐ Show Number Of Pages Only

(See the <u>CDX Server API Reference</u> for more advanced query options.)

<u>**Back To All Indexes**</u>

Figure 6-2. *Common crawl index search bar*

We'll get a JSON object as shown in Figure 6-3. Each dictionary represents one page capture in the common crawl index. It includes the status OK captures (status code 200) as well as redirects (301) and others. We also get additional details such as the mime type of the web page, timestamp of the capture, and so on. It's of particular interest to us to note down the filename, offset, and length of a particular record since that will allow us to fetch a particular HTML record from the S3 bucket shown a little later in this section.

```
{"urlkey": "com,apress)/", "timestamp": "20200330111944", "redirect": "https://www.apress.com/",
"status": "301", "url": "http://www.apress.com/", "mime": "unk", "digest":
"3I42H3S6NNFQ2MSVX7XZKYAYSCX5QBYJ", "offset": "5289148", "filename": "crawl-data/CC-MAIN-2020-16
/segments/1585370496901.28/crawldiagnostics/CC-MAIN-20200330085157-20200330115157-00114.warc.gz",
"length": "637", "mime-detected": "application/octet-stream"}
{"urlkey": "com,apress)/", "timestamp": "20200330111944", "redirect": "https://www.apress.com/us",
"status": "303", "url": "https://www.apress.com/", "mime": "unk", "digest":
"3I42H3S6NNFQ2MSVX7XZKYAYSCX5QBYJ", "offset": "18414115", "filename": "crawl-data/CC-MAIN-2020-16
/segments/1585370496901.28/crawldiagnostics/CC-MAIN-20200330085157-20200330115157-00499.warc.gz",
"length": "662", "mime-detected": "application/octet-stream"}
{"urlkey": "com,apress)/", "timestamp": "20200330174356", "redirect": "https://www.apress.com/",
"status": "301", "url": "http://www.apress.com/", "mime": "unk", "digest":
"3I42H3S6NNFQ2MSVX7XZKYAYSCX5QBYJ", "offset": "5036577", "filename": "crawl-data/CC-MAIN-2020-16
/segments/1585370497171.9/crawldiagnostics/CC-MAIN-20200330150913-20200330180913-00114.warc.gz",
"length": "636", "mime-detected": "application/octet-stream"}
{"urlkey": "com,apress)/", "timestamp": "20200330174356", "redirect": "https://www.apress.com/us",
"status": "303", "url": "https://www.apress.com/", "mime": "unk", "digest":
"3I42H3S6NNFQ2MSVX7XZKYAYSCX5QBYJ", "offset": "17864580", "filename": "crawl-data/CC-MAIN-2020-16
/segments/1585370497171.9/crawldiagnostics/CC-MAIN-20200330150913-20200330180913-00499.warc.gz",
"length": "663", "mime-detected": "application/octet-stream"}
{"urlkey": "com,apress)/", "timestamp": "20200408143519", "redirect": "https://www.apress.com/",
"status": "301", "url": "http://www.apress.com/", "mime": "unk", "digest":
"3I42H3S6NNFQ2MSVX7XZKYAYSCX5QBYJ", "offset": "5438315", "filename": "crawl-data/CC-MAIN-2020-16
/segments/1585371818008.97/crawldiagnostics/CC-MAIN-20200408135412-20200408165912-00114.warc.gz",
"length": "636", "mime-detected": "application/octet-stream"}
```

Figure 6-3. *Screenshot of the results page from the common crawl index*

Querying this GET API programmatically is pretty intuitive; just go to the address bar and you will get the parameters for this query as shown in the following:

```
http://index.commoncrawl.org/CC-MAIN-2020-16-index?url=apress.
com%2F*&output=json
```

We can filter the query to only show the status code of 200 and a mime-detected of text/html as shown in the following:

```
http://index.commoncrawl.org/CC-MAIN-2020-16-index?url=apress.
com%2F*&filter==mime-detected:text/html&filter==status:200&output=json
```

Lastly, we check the number of pages by using the following query. For a complete guide on how to paginate through the results and other filters, check out the documentation (`https://github.com/webrecorder/pywb/wiki/CDX-Server-API#api-reference`).

```
http://index.commoncrawl.org/CC-MAIN-2020-16-index?url=apress.com%2F*&showN
umPages=true&output=json
# Output

{"pages": 2, "pageSize": 5, "blocks": 9}
```

This API works pretty great for few and infrequent requests; however, if you are trying to use it for bulk querying for dozens of domain addresses, then I would have to warn you against that since it can easily bring down the server. Once you start throwing over half a dozen queries in a short period of time, the server starts throttling down your requests by sending a 503 service unavailable error. We will discuss Amazon Athena in Chapter 7, and it will be perfect for running batch queries on the common crawl index.

Let us parameterize the same URL we opened in Listing 6-3 using urllib and generate the GET request URL for the common crawl index as shown in Listing 6-4.

Listing 6-4. Parameterizing the URL for searching the common crawl index

```
import urllib

def get_index_url(query_url):

    query = urllib.parse.quote_plus(query_url)
    base_url = 'https://index.commoncrawl.org/CC-MAIN-2020-16-index?url='
    index_url = base_url + query + '&output=json'
    return index_url
```

```
query_url = 'http://archive.griffith.ox.ac.uk/index.php/informationobject/
browse?view=card&languages=en&creators=393&mediatypes=136&sort=referenceCod
e&sf_culture=en&levels=223&topLod=0&limit=30&sortDir=asc'
index_url = get_index_url(query_url)
print(index_url)
#Output
```

```
https://index.commoncrawl.org/CC-MAIN-2020-16-index?url=http%3A%2F%2Farchive.
griffith.ox.ac.uk%2Findex.php%2Finformationobject%2Fbrowse%3Fview%3Dcard%26
languages%3Den%26creators%3D393%26mediatypes%3D136%26sort%3DreferenceCode%2
6sf_culture%3Den%26levels%3D223%26topLod%3D0%26limit%3D30%26sortDir%3Dasc&o
utput=json
```

We can use the requests library to fetch the index JSON in Listing 6-5.

Listing 6-5. Getting results from the common crawl index

```
import re
import time
import gzip
import json
import requests
try:
    from io import BytesIO
except:
    from StringIO import StringIO
def get_index_json(index_url):

    payload_content = None

    for i in range(4):
        resp = requests.get(index_url)
        print(resp.status_code)

        time.sleep(0.2)

        if resp.status_code == 200:
```

```
for x in resp.content.strip().decode().split('\n'):
            payload_content = json.loads(x)
        break
    return payload_content
index_json = get_index_json(index_url)
print(index_json)
#Output
```

{'urlkey': 'uk,ac,ox,griffith,archive)/index.php/informationobject/
browse?creators=393&languages=en&levels=223&limit=30&mediatypes=136
&sf_culture=en&sort=referencecode&sortdir=asc&toplod=0&view=card',
'timestamp': '20200328100119', 'status': '200', 'url': 'http://archive.
griffith.ox.ac.uk/index.php/informationobject/browse?view=card&languag
es=en&creators=393&mediatypes=136&sort=referenceCode&sf_culture=en&lev
els=223&topLod=0&limit=30&sortDir=asc', 'mime': 'text/html', 'digest':
'LLZBM2KWPSEKOAK23C4J2V2FK5NLXNUC', 'charset': 'UTF-8', 'offset': '14692801',
'filename': 'crawl-data/CC-MAIN-2020-16/segments/1585370490497.6/warc/CC-MA
IN-20200328074047-20200328104047-00455.warc.gz', 'length': '6409', 'mime-
detected': 'text/html', 'languages': 'eng'}

As a sanity check, notice that the filename fetched from the preceding index JSON matches the filename in Listing 6-2.

Let's fetch the raw HTML page from the common crawl archives located on the S3 bucket. It's encoded as a WARC object. You can check out the entire file format standard referenced in the introduction, but individual warc objects are pretty simple to understand; they are just individually gzipped records which have combined crawl-related metadata, the response headers, and the raw HTML itself separated by the '\r\n\r\n'.

Listing 6-6 is using the offsets and size of the individual record from the common crawl index to fetch it and unzip the files to separate out the three components.

Listing 6-6. Getting web page data from the S3 bucket

```python
def get_from_index(page):

    offset, length = int(page['offset']), int(page['length'])
    offset_end = offset + length - 1
    prefix = 'https://commoncrawl.s3.amazonaws.com/'

    try:

        r = requests.get(prefix + page['filename'], headers={'Range':
        'bytes={}-{}'.format(offset, offset_end)})
        raw_data = BytesIO(r.content)
        f = gzip.GzipFile(fileobj=raw_data)
        data = f.read()

    except:

        print('some error in connection?')

    try:
        warc, header, response = data.strip().decode('utf-8').split('\r\n\
        r\n', 2)
    except Exception as e:
        pass
        print(e)

    return warc, header, response
warc, header, response = get_from_index(index_json)
```

The response headers and raw HTML shown in Listing 6-7 are exactly the same as Listing 6-3, what you would've gotten by iterating through the records of the entire warc file or requesting this web page via the requests library or browser.

Listing 6-7. Header and HTML

```python
print(header)
print('*'*10)
print(response)
```

```
#Output
HTTP/1.1 200 OK
Server: nginx/1.14.0 (Ubuntu)
Date: Sat, 28 Mar 2020 10:01:19 GMT
Content-Type: text/html; charset=utf-8
X-Crawler-Transfer-Encoding: chunked
Connection: keep-alive
Set-Cookie: symfony=jjecpro8lfekf6nmo9hj7qc5eb; path=/; HttpOnly
Expires: Thu, 19 Nov 1981 08:52:00 GMT
Cache-Control: no-store, no-cache, must-revalidate
Pragma: no-cache
X-Ua-Compatible: IE=edge,chrome=1
X-Crawler-Content-Encoding: gzip
Content-Length: 31469
*********
<!DOCTYPE html>
<html lang="en" dir="ltr">
  <head>
    <meta http-equiv="Content-Type" content="text/html; charset=utf-8" />
<meta http-equiv="X-Ua-Compatible" content="IE=edge,chrome=1" />
    <meta name="title" content="Griffith Institute Archive" />
. . .
html truncated
```

We learned about how to parse raw HTML using the BeautifulSoup library in Chapter 2; let's use it along with a regex-based preprocessor function in Listing 6-8 to extract text from this web page.

Listing 6-8. Extracting text from the web page

```
from bs4 import BeautifulSoup
import re
def preprocessor_final(text):
    if isinstance((text), (str)):
        text = re.sub('<[^>]*>', ' ', text)
        text = re.sub('[\W]+', ' ', text.lower())
        return text
```

```
    if isinstance((text), (list)):
        return_list = []
        for i in range(len(text)):
            temp_text = re.sub('<[^>]*>', '', text[i])
            temp_text = re.sub('[\W]+', '', temp_text.lower())
            return_list.append(temp_text)
        return(return_list)

soup = BeautifulSoup(response,'html.parser')
for script in soup(["script","style"]):
        script.extract()
print(preprocessor_final(soup.get_text()).replace('\n', ' '))

#Output
```

griffith institute archive griffith institute archive log in have an account email password log in quick links quick
links home about help global search replace privacy policy language language english français español nederlands port
uguês deutsch čeština clipboard clipboard clear all selectionsgo to clipboardload clipboardsave clipboard browse brow
se collectionspeople and organisationsplacessubjectsdigital objects search search advanced search filters narrow your
results by language unique records 1 results 1 english 1 results 1 creator all gardiner sir alan henderson 1 results
1 level of description all collection 1 results 1 media type all image 1 results 1 showing 1 results archival descrip
tion gardiner sir alan henderson collection image english advanced search options find results with and or not in any
field title archival history scope and content extent and medium subject access points name access points place acces
s points genre access points identifier reference code digital object text finding aid text creator any field except
finding aid text add new criteria and or not limit results to top level description filter results by level of descri
ption collection file fonds item object part series subfonds subseries digital object available yes no finding aid ye
s no generated uploaded copyright status public domain under copyright unknown general material designation architect
ural drawing cartographic material graphic material moving images multiple media object philatelic record sound recor
ding technical drawing textual record top level descriptions all descriptions filter by date range start end overlapp
ing exact use these options to specify how the date range returns results exact means that the start and end dates of
descriptions returned must fall entirely within the date range entered overlapping means that any description whose s
tart or end dates touch or overlap the target date range will be returned print preview view sort by reference code d
ate modified title relevance identifier start date end date direction ascending descending 1 results with digital obj
ects show results with digital objects add to clipboard alan henderson gardiner collection printed 2020 03 28

Once you have the raw HTML response as well as the full text of the page, you can use any of the methods described in Chapters 2 and 4, respectively, to extract additional information from it.

Let us put aside the applications of HTML parsing itself for Chapter 7 and focus our attention on the applications of text files from the common crawls.

WET file format

The Common Crawl Foundation extracts text from all the crawled web pages and saves it as a WET file. The naming scheme for a WET file is pretty intuitive; to get a WET file for an equivalent WARC file, we simply change the ".warc.gz" to ".warc.**wet**.gz" and insert "wet" in the file path instead of "warc" as shown in the following:

WARC file: crawl-data/CC-MAIN-2020-16/segments/1585370490497.6/**warc**/CC-MAIN
-20200328074047-20200328104047-00455.warc.gz

WET file: crawl-data/CC-MAIN-2020-16/segments/1585370490497.6/**wet**/CC-MAIN
-20200328074047-20200328104047-00455.warc.**wet**.gz

There are lots of obvious advantages with working with text files directly such as the drastic reduction in size of the overall corpus; the WET files are only about 8–9 TB in size vs. 62 TB for the raw web crawls packaged as WARC files.

It's perfect for many natural language processing–based approaches such as text clustering, topic modeling, and text classification. Unfortunately, there is no index available for WET files so we will have to iterate through the entire file to extract text for a specific web page as shown in Listing 6-9. Save the WET files with the warc.gz extension on your local computer or server so it can be processed by the warc package.

Listing 6-9. Processing WET files

```
from time import time
import warc

file_name = 'YOUR_LOCAL_FILEPATH.warc.gz'
wet_path = 'crawl-data/CC-MAIN-2020-16/segments/1585370490497.6/wet/CC-MAIN
-20200328074047-20200328104047-00455.warc.wet.gz'
import boto3

from botocore.handlers import disable_signing
resource = boto3.resource('s3')
resource.meta.client.meta.events.register('choose-signer.s3.*', disable_
signing)

bucket = resource.Bucket('commoncrawl')

resource.meta.client.download_file('commoncrawl', wet_path, file_name)

def process_wet(file_name, limit=100):
    warc_file = warc.open(file_name, 'rb')
    t0 = time()
    n_documents = 0
```

```
    url_list = []
    #header_list = []
    html_content = []

    for i, record in enumerate(warc_file):

        url = record.url
        payload = record.payload.read()

        if url is None or payload is None or payload == b'':

            continue

        else:
            try:

                html_content.append(preprocessor_final(payload.
                decode('utf-8')))
                url_list.append(url)

            except Exception as e:
                #print(e)
                continue

        n_documents += 1

    warc_file.close()
    print('Parsing took %s seconds and went through %s documents' %
(time() - t0, n_documents))
    return html_content, url_list

file_name = 'YOUR_LOCAL_FILEPATH.warc.gz'
html_content, url_list = process_wet(file_name, limit = 10000000)
# Output
Parsing took 44.381158113479614 seconds and went through 53271 documents
```

Let's check the full text for our URL in Listing 6-10 and make sure that it matches with the text in Listing 6-8 we extracted from the warc file directly.

Listing 6-10. Printing the WET record

```
index_no = url_list.index(query_url)
print(html_content[index_no])
```

```
'griffith institute archive griffith institute archive log in have an account email password log in quick links quick
links home about help global search replace privacy policy language language english français español nederlands port
uguês deutsch čeština clipboard clipboard clear all selections go to clipboard load clipboard save clipboard browse b
rowse collections people and organisations places subjects digital objects search search advanced search filters narr
ow your results by language unique records 1 results 1 english 1 results 1 creator all gardiner sir alan henderson 1
results 1 level of description all collection 1 results 1 media type all image 1 results 1 showing 1 results archival
description gardiner sir alan henderson collection image english advanced search options find results with and or not
in any field title archival history scope and content extent and medium subject access points name access points plac
e access points genre access points identifier reference code digital object text finding aid text creator any field
except finding aid text add new criteria and or not limit results to top level description filter results by level of
description collection file fonds item object part series subfonds subseries digital object available yes no finding
aid yes no generated uploaded copyright status public domain under copyright unknown general material designation arc
hitectural drawing cartographic material graphic material moving images multiple media object philatelic record sound
recording technical drawing textual record top level descriptions all descriptions filter by date range start end ove
rlapping exact use these options to specify how the date range returns results exact means that the start and end dat
es of descriptions returned must fall entirely within the date range entered overlapping means that any description w
hose start or end dates touch or overlap the target date range will be returned print preview view sort by reference
code date modified title relevance identifier start date end date direction ascending descending 1 results with digit
al objects show results with digital objects add to clipboard alan henderson gardiner collection printed 2020 03 28 '
```

Website similarity

Now that we have familiarized ourselves with the WET file format, let us use it for calculating the similarity between web pages.

Similarity scoring of web pages is used to create graphics such as Figure 1-8 in Chapter 1 which showed that Alexa considers apress.com's visitors and search keywords to be very similar to manning.com and few other results. It also showed an overlap score with each of the similar sites.

We will convert web page text into tf-idf vectors which we have seen in Chapter 4, and we will use them to calculate the cosine-based similarity using the nearest neighbor algorithm.

There are numerous applications for similarity scores between text documents; they can be used for content recommendation systems and semantic searching, which is simply searching for documents based on their overall meaning instead of more common keyword or lexical searching. They are also used as a crawling strategy in broad crawlers such as Apache Nutch (`https://cwiki.apache.org/confluence/display/NUTCH/SimilarityScoringFilter`). Plagiarism checkers also rely on text similarity to detect potential plagiarism between web pages.

We'll have to modify our process_wet function to only save documents in English as shown in Listing 6-11 so that we can directly vectorize them using tf-idf. We'll use a library called compact language detection (cld2, https://pypi.org/project/cld2-cffi/); it can be downloaded by pip install cld2-cffi.

Listing 6-11. Processing WET files

```python
from time import time
import cld2
import pandas as pd

import warc

def process_wet(file_name, limit=100):
    warc_file = warc.open(file_name, 'rb')
    t0 = time()
    n_documents = 0

    url_list = []
    #header_list = []
    html_content = []

    for i, record in enumerate(warc_file):

        url = record.url
        payload = record.payload.read()

        if url is None or payload is None or payload == b'':

            continue

        else:
            try:

                isReliable, textBytesFound, details = cld2.detect(payload.
                decode('utf-8'))

                lang1 = details[0][1]
                lang1_per = details[0][2]

                lang2 = details[1][1]
                lang2_per = details[1][2]
```

```python
        if lang1 == 'en' and lang1_per > 98 and lang2 == 'un' and
        len(str(payload).split(" ")) > 100:

            html_content.append(preprocessor_final(payload.
            decode('utf-8')))
            url_list.append(url)

    except Exception as e:
        #print(e)
        continue

  n_documents += 1

warc_file.close()
print('Parsing took %s seconds and went through %s documents' %
(time() - t0, n_documents))
return html_content, url_list

file_name = 'YOUR_LOCAL_FILEPATH.warc.gz'
html_content, url_list = process_wet(file_name, limit = 10000000)
df = pd.DataFrame({"full_text":html_content, "url":url_list})
df.head()
# Output
Parsing took 35.51691484451294 seconds and went through 52442 documents
```

	full_text	url
0	close up characters 94 game answers for 100 es...	http://100escaperswalkthrough.com/category/clo...
1	105 pymble house the most beautiful asian godd...	http://105pymblehouse.com.au/profiles.php?l=131
2	5 1080p lcd hdtv free shipping 5 1080p lcd hd...	http://1080plcdhdtvfreeshipping.blogspot.com/
3	presidential maroons 12 apostrophes digression...	http://12apostrophes.net/presidential-maroons/
4	link url rotator service promote all your prog...	http://1linkurl.com/

Let's vectorize the text using tf-idf vectorization in Listing 6-12 as seen in Chapter 4.

Listing 6-12. Vectorizing text

```
from sklearn.feature_extraction.text import TfidfVectorizer

tfidf_transformer = TfidfVectorizer(stop_words='english',
                                    ngram_range=(1, 2), lowercase=True, max_
                                    features=20000)

X_train_text = tfidf_transformer.fit_transform(df["full_text"])
df_dtm = pd.DataFrame(X_train_text.toarray(), columns=tfidf_transformer.
get_feature_names())
df_dtm.head()
```

	00	00 00	00 01	00 04	00 07	00 08	00 09	00 10	00 11	00 12	...	zombie	zombies	zone	zone pass	zone puck	zones	zoning	zoo	zoom	zte
0	0.0	0.0	0.0	0.0	0.0	0.0	0.0	0.0	0.0	0.0	...	0.0	0.0	0.0	0.0	0.0	0.0	0.0	0.000000	0.000000	0.0
1	0.0	0.0	0.0	0.0	0.0	0.0	0.0	0.0	0.0	0.0	...	0.0	0.0	0.0	0.0	0.0	0.0	0.0	0.000000	0.000000	0.0
2	0.0	0.0	0.0	0.0	0.0	0.0	0.0	0.0	0.0	0.0	...	0.0	0.0	0.0	0.0	0.0	0.0	0.0	0.000000	0.026344	0.0
3	0.0	0.0	0.0	0.0	0.0	0.0	0.0	0.0	0.0	0.0	...	0.0	0.0	0.0	0.0	0.0	0.0	0.0	0.045599	0.000000	0.0
4	0.0	0.0	0.0	0.0	0.0	0.0	0.0	0.0	0.0	0.0	...	0.0	0.0	0.0	0.0	0.0	0.0	0.0	0.000000	0.000000	0.0

5 rows × 20000 columns

Let's fit the nearest neighbor algorithm on these tf-idf vectors as shown in Listing 6-13.

Listing 6-13. Applying the nearest neighbor algorithm

```
from sklearn.neighbors import NearestNeighbors

NN= NearestNeighbors(n_neighbors=5, radius=1.0, algorithm='auto',
leaf_size=30, metric='cosine', p=2, metric_params=None, n_jobs=None)
NN.fit(X_train_text)
# Output
NearestNeighbors(algorithm='auto', leaf_size=30, metric='cosine',
                 metric_params=None, n_jobs=None, n_neighbors=5, p=2,
                 radius=1.0)
```

We will use the fitted nearest neighbor model to get the five nearest neighbors to the web page from Listing 6-14. It uses cosine distances as a measure of similarity between two vectors. The lower the distance between two vectors, the higher the similarity between them, within a range of 0 to 1.

It's a tuple of two arrays; the first array contains cosine distances of a vector with its neighbors, and the second array lists the index numbers of the neighbors. The first value shows a cosine distance of zero; that's unsurprising since the distance of a vector to itself is zero.

Listing 6-14. Calculating nearest neighbors

```
neigh_list = NN.kneighbors(df_dtm.iloc[134].values.reshape(1, -1),
n_neighbors=5, return_distance=True)
print(neigh_list)
#Output
(array([[0.        , 0.15182213, 0.16689516, 0.2082976 , 0.21517839]]),
array([[ 134, 4195,   133, 4125, 1631]], dtype=int64))
```

We can load this up in a pandas dataframe as shown in Listing 6-15.

Listing 6-15. Loading neighbors into a dataframe

```
neigh_df = pd.DataFrame({"url_index":neigh_list[1][0].tolist(), "cosine_
dist":neigh_list[0][0].tolist()})
neigh_df.head()
# Output
```

	cosine_dist	url_index
0	0.000000	134
1	0.151822	4195
2	0.166895	133
3	0.208298	4125
4	0.215178	1631

We will print the URLs of these similar pages in Listing 6-16 and the full text of one of the similar pages.

Listing 6-16. Printing nearest neighbor text and URLs

```
for i in range(len(neigh_df)):
    print(url_list[neigh_df.url_index.iloc[i]])
    print("*"*10)
```

```
http://archive.griffith.ox.ac.uk/index.php/informationobject/browse?view=
card&languages=en&creators=393&mediatypes=136&sort=referenceCode&sf_culture=
en&levels=223&topLod=0&limit=30&sortDir=asc
**********
https://atom.library.yorku.ca/index.php/informationobject/browse?places=556
037&view=card&subjects=558646&sort=identifier&sf_culture=fi&%3Bview=card&%3
Bsort=alphabetic&sortDir=asc
**********
http://archive.griffith.ox.ac.uk/index.php/informationobject/browse?sf_ultu
re=cs&creators=9811&sortDir=desc&sort=lastUpdated&%3Bsort=lastUpdated&%
3Bnames=21267&%3Blevels=223&%3BtopLod=0&%3Blimit=30
**********
https://archives.jewishmuseum.ca/informationobject/browse?sortDir=desc&crea
tors=110543&levels=221&%3Bsubjects=400&%3Bamp%3Bsort=relevance&%3Bsort=
alphabetic&sort=alphabetic
**********
http://rbscarchives.library.ubc.ca/index.php/informationobject/browse?sort=
lastUpdated&places=555934%2C555933%2C555931&names=555848%2C555877%2C555869%
2C555870&%3Bamp%3Bcollection=183468&%3Bamp%3Bview=card&%3Bamp%3BonlyMedia=
1&%3Bamp%3BtopLod=0&%3Bamp%3Bsort=alphabetic&%3Bsort=alphabetic
**********
print(html_content[4195])
```

'york university libraries clara thomas archives special collections ok this website uses cookies to enhance your abi
lity to browse and load content log in have an account email password log in quick links quick links home about help
privacy policy clipboard clipboard clear all selections go to clipboard load clipboard save clipboard browse browse a
rchival descriptions people and organizations subjects places digital objects search search global search advanced se
arch filters language unique records 1 results 1 englanti 1 results 1 place all toronto 1 results 1 yellowknife n w t
1 results 1 quebec 1 results 1 united states 1 results 1 canada 1 results 1 ontario 1 results 1 new york 1 results 1
new york city 1 results 1 subject all business and commerce 1 results 1 mining 1 results 1 gold 1 results 1 natural r
esources 1 results 1 showing 1 results archival description only top level descriptions new york city mining advanced
search options find results with and or not in any field title scope and content extent and medium subject access poi
nts name access points place access points genre access points identifier reference code digital object text finding
aid text creator any field except finding aid text add new criteria and or not limit results to repository http viaf
org york university archives special collections ctasc top level description filter results by level of description a
ccession collection file fonds item series digital object available yes no finding aid yes no generated uploaded gene
ral material designation architectural drawing cartographic material graphic material moving images object philatelic
record sound recording technical drawing textual record top level descriptions all descriptions filter by date range
start end overlapping exact use these options to specify how the date range returns results exact means that the star
t and end dates of descriptions returned must fall entirely within the date range entered overlapping means that any
description whose start or end dates touch or overlap the target date range will be returned print preview view sort
by identifier date modified title relevance reference code start date end date direction ascending descending arnold
hoffman fonds add to clipboard arnold hoffman fonds printed 2020 03 28 '

It's apparent just by looking at the URLs that all the other similar pages are also from online library archives of major educational and nonprofit institutions.

A web page–level similarity for just one page is too little to make a definitive judgment about whether the domain "`http://archive.griffith.ox.ac.uk`" is similar to "`https://atom.library.yorku.ca`" like we have seen with Alexa website similarity diagrams.

We can create a SQL database to save web page similarity scores by iterating through all the available WET files from a monthly crawl and use this database to query for top domains with the highest number of similar web pages.

You'll undoubtedly run into memory issues once the number of WET files processed exceeds a few hundred depending on your local machine.

It may be a good idea to restrict the number of text from a given web page to only the first few lines. Another common filtering technique is to only process web pages which are one or two levels away from the root or base domain. Many similar site databases only consider text enclosed by the meta description and title tags, and you'll learn how to get that without HTML parsing in the next section. Even with these modifications, you are looking at vectorizing about 20–30 GB of text if you process all WET files from a monthly crawl.

You can vectorize text using a more memory-efficient algorithm like sklearn's hashing vectorizer (`https://scikit-learn.org/stable/modules/generated/sklearn.feature_extraction.text.HashingVectorizer.html`). I also recommend switching to an approximate nearest neighbor algorithm such as FLANN (Fast Library for Approximate Nearest Neighbors, `https://github.com/mariusmuja/flann`).

WAT file format

The Common Crawl Foundation parses all the metadata associated with a web page such as HTTP request and response headers, outgoing links, meta tags from a web page, and so on and saves them as a JSON into a separate file with a WAT file extension.

Their total size is about 20 TB for each monthly crawl vs. ~62 TB for an equivalent WARC file. The naming scheme for a WAT file is similar to a WET file; we simply change ".warc.gz" to ".warc.**wat**.gz" and insert "wat" in the file path instead of "warc". There is no index available for directly fetching a record by querying a URL so we have to iterate through the entire WAT file. Download and save the WAT file with a "warc.gz" extension so that we can process it using the warc package as shown in Listing 6-17.

Listing 6-17. Processing and iterating through WAT files

```
# NOTE: we got this wat_path from going to https://commoncrawl.org/2020/04/
march-april-2020-crawl-archive-now-available/

wat_path = 'crawl-data/CC-MAIN-2020-16/segments/1585370490497.6/wat/CC-MAIN
-20200328074047-20200328104047-00000.warc.wat.gz'
file_name = 'YOUR_LOCAL_FILEPATH.warc.gz'
import boto3

from botocore.handlers import disable_signing
resource = boto3.resource('s3')
resource.meta.client.meta.events.register('choose-signer.s3.*', disable_
signing)

bucket = resource.Bucket('commoncrawl')

resource.meta.client.download_file('commoncrawl', wat_path, file_name)

from time import time
import warc

def process_wat(file_name, limit=10000):
    warc_file = warc.open(file_name, 'rb')
    t0 = time()
    n_documents = 0
```

```
    url_list = []
    header_list = []
    html_content = []

    for i, record in enumerate(warc_file):

        if n_documents >= limit:

            break

        url = record.url
        payload = record.payload.read()
        html_content.append(payload)
        url_list.append(url)

        n_documents += 1

    warc_file.close()
    print('Parsing took %s seconds and went through %s documents' %
    (time() - t0, n_documents))
    return html_content, url_list
file_name = 'YOUR_LOCAL_FILEPATH.warc.gz'
html_content, url_list = process_wat(file_name, limit = 1000000)
#output
Parsing took 20.070790767669678 seconds and went through 160415 documents
```

WAT files consist of individual records from web pages each in a JSON object. In this case, we simply loaded them onto a list.

Let us go through one entry at random and see all the individual information which is packaged as a WAT record in Listing 6-18.

Listing 6-18. Exploring the WAT record

```
import json
sample_dict = json.loads(html_content[60000])
sample_dict
# Output
```

{'**Container**': {'Compressed': True,
  'Filename': 'CC-MAIN-20200328074047-20200328104047-00000.warc.gz',
  'Gzip-Metadata': {'Deflate-Length': '7053',
   'Footer-Length': '8',
   'Header-Length': '10',
   'Inflated-CRC': '489333750',
   'Inflated-Length': '28943'},
  'Offset': '365278980'},
 '**Envelope**': {'Format': 'WARC',
  '**Payload-Metadata**': {'Actual-Content-Length': '28338',
   'Actual-Content-Type': 'application/http; msgtype=response',
   'Block-Digest': 'sha1:XCUMWTUZQSM3TGFOTW3F7CJKXORQBVG7',
   'HTTP-Response-Metadata': {'Entity-Digest': 'sha1:2NBB2Q47ZGHHSNHIXQWMEH
LTDNGVNTSD',
    'Entity-Length': '28109',
    'Entity-Trailing-Slop-Length': '0',
    '**HTML-Metadata**': {'**Head**': {'Link': [{'path': 'LINK@/href',
        'rel': 'icon',
        'url': '../site//img/favicon.png'},
       {'path': 'LINK@/href',
        'rel': 'canonical',
        'url': 'https://bestsports.com.br/bi/atlbihome.php?esp=42'},
       {'path': 'LINK@/href',
        'rel': 'alternate',
        'url': 'https://bestsports.com.br/bi/atlbihome.php?esp=42'},
       {'path': 'LINK@/href',
        'rel': 'alternate',
        'url': 'https://bestsports.com.br/bi/atlbihome.php?esp=42&lang=2'},
       {'path': 'LINK@/href',
        'rel': 'alternate',
        'url': 'https://bestsports.com.br/bi/atlbihome.php?esp=42&lang=3'},
       {'path': 'LINK@/href',
        'rel': 'alternate',
        'url': 'https://bestsports.com.br/bi/atlbihome.php?esp=42&lang=4'},
       {'path': 'LINK@/href',

```
      'rel': 'alternate',
      'url': 'https://bestsports.com.br/bi/atlbihome.php?esp=42&lang=2'},
   {'path': 'LINK@/href',
    'rel': 'stylesheet',
    'type': 'text/css',
    'url': './img/BSbi2019.css'},
   {'path': 'LINK@/href',
    'rel': 'stylesheet',
    'type': 'text/css',
    'url': '../db/img/BSdb2019.css'},
   {'path': 'LINK@/href',
    'rel': 'stylesheet',
    'type': 'text/css',
    'url': '../site/img/BSsite2019.css'}],
 'Metas': [{'content': 'width=device-width, initial-scale=1.0',
    'name': 'viewport'},
   {'content': 'no-cache, no-store', 'http-equiv': 'Cache-Control'},
   {'content': 'no-cache, no-store', 'http-equiv': 'Pragma'},
   {'content': 'eb96de70-e940-11e9-b21b-d1121cb6cef8',
    'name': 'axl-verification'},
   {'content': 'pt-BR', 'http-equiv': 'Content-Language'},
   {'content': '0', 'http-equiv': 'Expires'}],
 'Scripts': [{'path': 'SCRIPT@/src',
    'url': '//pagead2.googlesyndication.com/pagead/js/adsbygoogle.js'},
   {'path': 'SCRIPT@/src', 'url': 'https://d3js.org/d3.v4.min.js'},
   {'path': 'SCRIPT@/src',
    'url': 'https://cdnjs.cloudflare.com/ajax/libs/d3-tip/0.7.1/d3-tip.
    min.js'},
   {'path': 'SCRIPT@/src', 'url': './tools/biTools2019.js'}],
 'Title': 'BEST sports Analytics\ufeff - Atletas - Tiro Esportivo'},
 'Links': [{'path': 'A@/href', 'text': 'Doar', 'url': '../db/donate.
php'},
 {'alt': 'BESTsports Database',
  'path': 'IMG@/src',
  'title': 'BESTsports Database',
```

```
              'url': '../site/img/logoBarraDB.png'},
         {'path': 'A@/href', 'text': 'Database', 'url': '../db/homedb.php'},
         {'alt': 'BESTsports Analytics',
          'path': 'IMG@/src',
          'title': 'BESTsports Analytics',
          'url': '../site/img/logoBarraBI.png'},
          .
        . (entries truncated)
          .
         {'path': 'A@/href', 'text': 'Doar', 'url': '../db/donate.php'}]},
     'Headers': {'Connection': 'Keep-Alive',
       'Content-Length': '28109',
       'Content-Type': 'text/html; charset=UTF-8',
       'Date': 'Sat, 28 Mar 2020 08:49:45 GMT',
       'Keep-Alive': 'timeout=5, max=100',
       'Server': 'Apache',
       'X-Crawler-Transfer-Encoding': 'chunked'},
      'Headers-Length': '229',
      'Response-Message': {'Reason': 'OK',
       'Status': '200',
       'Version': 'HTTP/1.1'}},
    'Trailing-Slop-Length': '4'},
   'WARC-Header-Length': '601',
   'WARC-Header-Metadata': {'Content-Length': '28338',
    'Content-Type': 'application/http; msgtype=response',
    'WARC-Block-Digest': 'sha1:XCUMWTUZQSM3TGFOTW3F7CJKXORQBVG7',
    'WARC-Concurrent-To': '<urn:uuid:373d6376-b317-4599-919f-a2845fd9aa4f>',
    'WARC-Date': '2020-03-28T08:49:46Z',
    'WARC-IP-Address': '162.254.149.193',
    'WARC-Identified-Payload-Type': 'text/html',
    'WARC-Payload-Digest': 'sha1:2NBB2Q47ZGHHSNHIXQWMEHLTDNGVNTSD',
    'WARC-Record-ID': '<urn:uuid:cf047b12-2bff-4efc-ac77-7cb222bb96a4>',
    'WARC-Target-URI': 'https://bestsports.com.br/bi/atlbihome.php?esp=42',
    'WARC-Type': 'response',
    'WARC-Warcinfo-ID': '<urn:uuid:9574723b-0801-4dba-9d23-d3478ec93784>'}}}
```

We can see that each record is formatted as a deeply nested JSON file, with information from the HTML page enclosed in HTML metadata keys. It contains two keys within it: the Head key which contains all the information found within the <head>... </head> tags and the Links key which includes all the hyperlinks and anchor texts found on the page.

All the information found in an individual WAT record can be generated by simple HTML-based parsing libraries such as BeautifulSoup or lxml which we discussed in Chapter 2. For example, extracting information within meta tags, scripts, or hyperlinks can be done in a few lines of code; however, actually running it over a million of web pages will cost a lot in compute time so it's better to save the cost of parsing this ourselves and use this structured JSON file to better understand what this data can do for you.

```
sample_dict["Envelope"]['Payload-Metadata']['HTTP-Response-Metadata']
['HTML-Metadata'].keys()
# Output
dict_keys(['Head', 'Links'])
```
The Head typically includes links, meta tags, title tags, and scripts.

```
sample_dict["Envelope"]['Payload-Metadata']['HTTP-Response-Metadata']
['HTML-Metadata']["Head"].keys()
#Output

dict_keys(['Metas', 'Link', 'Title', 'Scripts'])
```

We can see which external JavaScript libraries are being used on the page.

```
sample_dict["Envelope"]['Payload-Metadata']['HTTP-Response-Metadata']
['HTML-Metadata']["Head"]["Scripts"]

#Output

[{'path': 'SCRIPT@/src',
  'url': '//pagead2.googlesyndication.com/pagead/js/adsbygoogle.js'},
 {'path': 'SCRIPT@/src', 'url': 'https://d3js.org/d3.v4.min.js'},
 {'path': 'SCRIPT@/src',
  'url': 'https://cdnjs.cloudflare.com/ajax/libs/d3-tip/0.7.1/d3-tip.min.js'},
 {'path': 'SCRIPT@/src', 'url': './tools/biTools2019.js'}]
```

Meta tags information such as shown is extracted and enclosed as a list.

```
<meta name="viewport" content="width=device-width, initial-scale=1.0">
<meta http-equiv="Cache-Control" content="no-cache, no-store" />
<meta http-equiv="Pragma" content="no-cache, no-store" />
<meta name="axl-verification" content="eb96de70-e940-11e9-b21b-
d1121cb6cef8">
```

```
sample_dict["Envelope"]['Payload-Metadata']['HTTP-Response-Metadata']
['HTML-Metadata']["Head"]["Metas"]
#Output
```

```
[{'content': 'width=device-width, initial-scale=1.0', 'name': 'viewport'},
 {'content': 'no-cache, no-store', 'http-equiv': 'Cache-Control'},
 {'content': 'no-cache, no-store', 'http-equiv': 'Pragma'},
 {'content': 'eb96de70-e940-11e9-b21b-d1121cb6cef8',
  'name': 'axl-verification'},
 {'content': 'pt-BR', 'http-equiv': 'Content-Language'},
 {'content': '0', 'http-equiv': 'Expires'}]
```
Similarly, information from title tags are available in the Title key.

```
<title>BEST sports
 Analytics - Atletas - Tiro Esportivo</title>
```

```
sample_dict["Envelope"]['Payload-Metadata']['HTTP-Response-Metadata']
['HTML-Metadata']["Head"]["Title"]
#Output
```

```
'BEST sports Analytics\ufeff - Atletas - Tiro Esportivo'
```

The main portion of a WAT file record is contained within the Links key where it contains all the extracted hyperlinks and anchor texts found on the page. In this case, this list contains 64 entries.

```
len(sample_dict["Envelope"]['Payload-Metadata']['HTTP-Response-Metadata']
['HTML-Metadata']["Links"])
#Output
64
```

Lastly, we also have the response header information saved under the Headers key which contains typical information such as servers, cache, compression type, and so on as shown in the following:

```
sample_dict["Envelope"]['Payload-Metadata']['HTTP-Response-Metadata']
["Headers"]
```

```
{'Connection': 'Keep-Alive',
 'Content-Length': '28109',
 'Content-Type': 'text/html; charset=UTF-8',
 'Date': 'Sat, 28 Mar 2020 08:49:45 GMT',
 'Keep-Alive': 'timeout=5, max=100',
 'Server': 'Apache',
 'X-Crawler-Transfer-Encoding': 'chunked'}
```

Now that we have a good idea of individual components of each WAT record, let us dive into some real-world applications for it in the next couple of sections.

Web technology profiler

Let's recall our discussion in Chapter 1 about web technology databases such as builtwith.com which collect a database of technologies used on web pages by web crawling.

A significant portion of this information such as server types, external JavaScript libraries used, font and styling, third-party verification services information, and so on comes from either the response headers or the information enclosed within <head>..</head> tags, both of which are extracted by WAT files and available for ingesting in a database quite readily.

You may be wondering about practical utilities of creating a database to check what technologies are used on a particular website. While knowing that a particular website uses a popular JavaScript library such as jQuery or MathJax might just be intellectual curiosity, it's of legitimate business interest to spot signs of paid widgets or apps which might in turn serve as the lead generation if your product is a potential competitor or even a stock market trading signal.

An interesting case study was reported by builtwith.com (`https://blog.builtwith.com/2018/07/17/web-focused-publicly-traded-tech-companies/`) where they showed a good correlation between the adoption of a widget as spotted by web crawling of a publicly traded company and its stock market price.

In almost all cases, such databases are powered by performing a regex search through the HTML source code and response headers for code snippets of interest which gives us a clue about the presence of a certain tool in the website's technology stack.

Listing 6-19 shows a regex search to see if the current web page uses Apache or not. Don't get too caught up in trying to understand the regex itself; all we are trying to do is search for the term "Apache" in a particular WAT record.

Listing 6-19. Regex search for the Apache server

```
import json
import re
sample_dict = json.loads(html_content[60000])
sample_dict
import time
x = re.compile("(?:Apache(?:$|/([\\d.]+)|[^/-]))|(?:^|\\b)HTTPD)").
search(str(sample_dict))
if x:
    print("This webpage uses Apache server")
else:
    print("no match found")
# Output
This website uses Apache server
```

There are a couple of optimizations we can make to get a more accurate and faster reading. You can use a faster regex engine like re2 discussed in Chapter 4 for email matching which can result in multifold improvement.

We can also restrict a regex search to only those areas of an HTML record where we are most likely to find a potential match as shown in Listing 6-20. In this example, server-related information is typically only contained in HTML response headers which can be accessed by the following object of the WAT record. It would make our regex search go considerably faster if we only search headers instead of the entire document.

Listing 6-20. Comparing total times for 1000 iterations

```
import json
import re
sample_dict = json.loads(html_content[60000])
sample_dict
import time
start_time = time.time()

for i in range(1000):
    x = re.compile("(?:Apache(?:$|/([\\d.]+)|[^/-])|(?:^|\\b)HTTPD)").
search(str(sample_dict))
end_time = time.time()
print("total time (for 1000 iterations) to check entire wat record: ", end_
time-start_time)
start_time = time.time()
for i in range(1000):
    x = re.compile("(?:Apache(?:$|/([\\d.]+)|[^/-])|(?:^|\\b)HTTPD)").
search(str(sample_dict["Envelope"]['Payload-Metadata']['HTTP-Response-
Metadata']["Headers"]))
end_time = time.time()
print("total time (for 1000 iterations) to check entire wat record header:
", end_time-start_time)
# Output
total time (for 1000 iterations) to check entire wat
record:  0.18289828300476074
total time (for 1000 iterations) to check entire wat record
header:  0.016824007034301758
```

Similarly, if we are looking to search for a string via regex enclosed within the script tag, then this can be achieved using a WAT record much faster, as shown in Listing 6-21, than trying to parse an entire HTML record.

Listing 6-21. Checking for Google AdSense in WAT records

```
x = re.compile("googlesyndication\\.com/").search(str(sample_
dict["Envelope"]['Payload-Metadata']['HTTP-Response-Metadata']['HTML-
Metadata']["Head"]["Scripts"]
))
if x:
    print("This website uses Google Adsense")

else:
    print("no match found")

#Output
This website uses Google Adsense
```

We could write the preceding regexes since we already knew that the web page used Google AdSense and Apache from visual inspection of WAT files in earlier sections. For unknown web pages, we can use a lookup table from a popular open source JavaScript library called Wappalyzer (https://github.com/AliasIO/wappalyzer/blob/master/src/apps.json) shown in Listing 6-22.

Listing 6-22. Code excerpt from the Wappalyzer library

```
"Apache": {
     "cats": [
       22
     ],
     "cpe": "cpe:/a:apache:http_server",
     "headers": {
       "Server": "(?:Apache(?:$|/([\\d.]+)|[^/-])|(?:^|\\b)
       HTTPD)\\;version:\\1"
     },
     "icon": "Apache.svg",
     "website": "http://apache.org"
},

"Varnish": {
     "cats": [
       23
```

```
  ],
  "headers": {
    "Via": "varnish(?: \\(Varnish/([\\d.]+)\\))?\\;version:\\1",
    "X-Varnish": "",
    "X-Varnish-Action": "",
    "X-Varnish-Age": "",
    "X-Varnish-Cache": "",
    "X-Varnish-Hostname": ""
  },
  "icon": "Varnish.svg",
  "website": "http://www.varnish-cache.org"
```

Wappalyzer also searches the HTML source and scripts for Google AdSense as shown in Listing 6-23.

Listing 6-23. Code excerpt from the Wappalyzer library for Google AdSense

```
"Google AdSense": {
  "cats": [
    36
  ],
  "icon": "Google AdSense.svg",
  "js": {
    "Goog_AdSense_": "",
    "__google_ad_urls": "",
    "google_ad_": ""
  },
  "script": [
    "googlesyndication\\.com/",
    "ad\\.ca\\.doubleclick\\.net",
    "2mdn\\.net",
    "ad\\.ca\\.doubleclick\\.net"
  ],
  "website": "https://www.google.fr/adsense/start/"
}
```

Once there is a regex match, it will package results as a dictionary where the keys will be category names, and values will be in the form of a list of technologies found for a particular category shown in Listing 6-24. There are lots of categories in the entire JSON file, but we are only showing a few of them.

Listing 6-24. Categories from the Wappalyzer library JSON

```
"22": {
        "name": "Web servers",
        "priority": 8
    },
"23": {
        "name": "Caching",
        "priority": 7
},
.
.
"36": {
        "name": "Advertising",
        "priority": 9
    },
```

There is an open source Python–based library called builtwith (https://pypi.org/project/builtwith/) (unrelated and not to be confused with builtwith.com) which uses the same Wappalyzer list shown earlier.

In Listing 6-25, we are using the builtwith package to identify technologies from the WAT response.

Listing 6-25. Using the builtwith library

```
import builtwith
import time
start_time = time.time()
print(builtwith.builtwith(url = 'none', html = html_content[60000], headers =
sample_dict["Envelope"]['Payload-Metadata']['HTTP-Response-Metadata']
["Headers"]))
end_time = time.time()
```

```
print(end_time-start_time)
#Output
{'web-servers': ['Apache'], 'advertising-networks': ['Google AdSense']}
0.5603644847869873
```

The real challenge for developing a commercially viable technology profiler database is not only to perform these regex scans on scale and load them onto a database but also to keep the technology lookup list updated which is what powers the library.

Almost all technology data providers out there including Specrom Analytics use their own proprietary technology lookup list which is manually curated pretty frequently by a team of front-end developers. In our case, we specifically provide technology data to Internet security researchers for the presence of malicious code fragments in broader web crawls; hence, our technology lookup list looks quite different than open source libraries such as Wappalyzer, but the general idea remains the same.

All commercial providers of web technology databases will augment scraped data with other sources. builtwith.com doesn't just rely on searching on HTTP response headers and HTML source pages to populate its database. They also pull information from DNS records themselves which give additional information such as email servers, hostnames, content delivery networks, and so on.

For example, builtwith.com shows that apress.com uses Varnish and Apache for web servers which we might already know from HTTP response headers, but it also tells us that it Fastly, Akamai for hosting, and a bunch of other data which we couldn't have possibly seen it from an HTML source page.

You can get the same type of information as shown in Listing 6-26 from any number of DNS lookup APIs such as Specrom's API on Algorithmia (`https://algorithmia.com/algorithms/specrom/DNS_Lookup`).

There is a treasure trove of information in DNS's TXT records which is frequently used to verify domain ownership for services such as DocuSign, Adobe Enterprise products, and so on.

For example, in the case of apress.com, we see that they have verified for Trustpilot, DocuSign, and GlobalSign as part of their TXT records. We would not have known that at all if we only relied on extracting information from scraped HTML pages.

Listing 6-26. DNS records information for apress.com

```
{
  "A": [
    "195.128.8.134"
  ],
  "CNAME": [
    "prod.springer.map.fastlylb.net."
  ],
  "MX": [
    "mxa-002c5801.gslb.pphosted.com.",
    "mxb-002c5801.gslb.pphosted.com."
  ],
  "NS": [
    "pdns5.ultradns.info.",
    "pdns6.ultradns.co.uk.",
    "pdns1.ultradns.net.",
    "pdns2.ultradns.net.",
    "pdns3.ultradns.org.",
    "pdns4.ultradns.org."
  ],
  "SOA": [
    "pdns1.ultradns.net. hostmaster.springer.com. 2016100551 14400 3600
    604800 86400"
  ],
  "TXT": [
    "\"CEOS1609226345\"",
    "\"docusign=c9615a9b-74c2-4d97-89d2-9955f478b3ab\"",
    "\"facebook-domain-verification=cbeonotfuvc96obohqqjav51vxpm2k\"",
    "\"google-site-verification=WfbC81vsslK5ANg-hlgcWcswtkFRcq-
    v9e4j4Ceh27A\"",
    "\"google-site-verification=tIroc0YgZ81OcoZA_ssClAruV-sVeytF0016-
    0qdVyM\"",
    "\"_globalsign-domain-verification=aQIRcZ4Sa1SmXZlbSPt5QrVI6ozleqaOsP
    LU-bElLs\"",
```

```
    "\"adobe-idp-site-verification=a4583e5d5c183b981fb8e3e44cfcd3835fa74b04
    b382b961ce05439b6d3a042d\"",
    "\"v=spf1 ip4:207.97.243.208 ip4:52.43.154.216 ip4:192.174.90.93
    ip4:52.41.1.125 ip4:192.174.90.94 include:_spf.google.com
    include:trustpilotservice.com include:sparkpostmail.com include:spf-
    002c5801.pphosted.com ~all\"",
    "\"MS=ms47091852\""
  ]
}
```

I just mentioned DNS records to make sure that even though we are talking about web scraping, we are not getting pigeonholed into thinking that it's the only way to get any structured information. DNS records are one important source of information which is not directly accessible from web scraping.

Backlinks database

We mentioned in Chapter 1 about the importance of backlinks in boosting a particular domain's trust authority from a search engine optimization (SEO) perspective. Comprehensive backlinks databases from commercial providers such as Ahrefs cost hundreds of dollars a month; we will show you how to get the same type of database with a little less coverage but only cost being computing power and storage.

Backlinks are also known as inlinks and are simply all the external links pointing you to the web page URLs at your domain from external domains.

Businesses want to know about backlinks to their domains and/or their competitors to know about the effectiveness of their content marketing programs to drive organic search engine–based traffic. They also want to check about how their or their competitors' products/services are being referenced by external sources such as news, blogs, and so on. Ahrefs, Moz, Alexa, and so on package this data to mainly serve this target audience.

However, this is not the only reason for calculating the number of backlinks to a page or domain. We need it for calculating domain importance which can be used to determine the crawl frequency and number of pages to be crawled from a particular domain; both of which are very vital if you are running a broad crawler, that is, a crawler similar to common crawl, Google, Bing, and so on, which isn't restricted to only fetching pages from a targeted set of web domains.

The idea of backlinks to measure a web page URL's overall importance is simple; if a higher number of pages reference a particular page, then it can be viewed as an implicit recommendation of the quality of the content found on that page, and consequently, web pages with higher number of backlinks should be ranked higher in search engine results than the ones with lower backlinks with everything else being equal. We can extend this logic to an overall domain-wide scale by calculating total referring domains to a particular domain and use that to calculate domain authority using algorithms such as PageRank or harmonic centrality.

WAT records contain hyperlinks pointing from a particular web page URL to other pages; this is referred to as outlinks.

We can use these outlinks to create a database for inlinks or backlinks by loading up the data in SQLite. In this example, we will only be doing it for all the hyperlinks found from one compressed WAT file, so it will be pretty doable; however, if you try to do it for all the WAT files for a monthly crawl, you will quickly realize the shortcomings of using SQLite due to very high memory requirements and long query times, and at that scale, it will be much better to switch to a graph database.

Let us iterate through the WAT file in Listing 6-27 and create a dataframe which can be loaded into a SQLite database.

Listing 6-27. Processing WAT to create a dataframe for the SQLite staging table

```
from time import time
import tld
import warc
import json

def process_wat_with_processing(file_name, limit=10000):
    warc_file = warc.open(file_name, 'rb')
    t0 = time()
    n_documents = 0

    url_list = []
    header_content_list = []
    html_content_list = []
    final_list = []
```

```python
    for i, record in enumerate(warc_file):
        #print(i)
        if n_documents >= limit:

            break

        url = record.url
        payload = record.payload.read()
        temp_dict = {}
        try:
            temp_dict["url"] = url
            temp_dict["webpage_source"] = tld.get_fld(url)
            #temp_dict['url_anchor_source'] = 'none'
            sample_dict = json.loads(payload)
            doc_links = sample_dict['Envelope']['Payload-Metadata']['HTTP-
            Response-Metadata']['HTML-Metadata']["Links"]
            for doc in doc_links:
                if doc["path"] == 'A@/href' and 'http' in doc["url"]:
                    temp_dict["backlink_source"] = tld.get_fld(doc["url"])
                    temp_dict["backlink"] = doc["url"]

                    #temp_dict["anchor_text"] = doc.get('text', 'none')
                    final_list.append(temp_dict.copy())

        except Exception as E:
            #print(E)
            continue

        n_documents += 1

    warc_file.close()
    print('Parsing took %s seconds and went through %s documents' %
(time() - t0, n_documents))
    return final_list

file_name = 'YOUR_LOCAL_FILEPATH.warc.gz'
final_list = process_wat_with_processing(file_name, limit = 100000)
```

```
import numpy as np
import pandas as pd

df = pd.DataFrame(final_list)
df.head()
```

	Backlink	backlink_source	url	webpage_source
0	http://000ojfb. wcomhost.com/ushwa/	wcomhost.com	http://000ojfb. wcomhost.com/ ushwa/2018-dan-pat...	wcomhost.com
1	https://www. facebook.com/ USHarnessWriters	facebook.com	http://000ojfb. wcomhost.com/ ushwa/2018-dan-pat...	wcomhost.com
2	https://aboutme. google.com/b/ 10733287508516707...	google.com	http://000ojfb. wcomhost.com/ ushwa/2018-dan-pat...	wcomhost.com
3	https://twitter.com/ USHWA_NATL	twitter.com	http://000ojfb. wcomhost.com/ ushwa/2018-dan-pat...	wcomhost.com
4	https://www. linkedin.com/ company-beta/ 25009386/	linkedin.com	http://000ojfb. wcomhost.com/ ushwa/2018-dan-pat...	wcomhost.com

Once we have the dataframe ready, we will create an empty SQLite database and insert data from the dataframe in Listing 6-28. It's an extremely straightforward schema, with three tables. The first one is the sources table which contains domain-specific information. The second table contains the web page URL and source_id. Lastly, we have a table called the backlinks table which maps a given webpage_id to its backlink_id.

Listing 6-28. Creating and inserting data in the SQLite backlinks database

```
from sqlalchemy import create_engine

engine = create_engine(r'sqlite:///sqlite_db_path', echo=True)
conn = engine.connect()

from sqlalchemy import Table, Column,UniqueConstraint, Integer, String,
DateTime, MetaData, ForeignKey

metadata = MetaData()

sources = Table('sources', metadata,
    Column('source_id', Integer, primary_key=True),
    Column('source_name', String, unique=True),
    Column('source_url', String, unique=True),
    Column('source_description', String)
    )

webpages = Table('webpages', metadata,
    Column('webpage_id', Integer, primary_key=True),
    Column('webpage_url', String, unique=True),
    Column('source_id', None, ForeignKey('sources.source_id')),
    )

backlinks = Table('backlinks', metadata,
    Column('backlink_id', Integer, primary_key=True),
    Column('link_id', None, ForeignKey('webpages.webpage_id')),
    Column('webpage_id', Integer),
    UniqueConstraint('webpage_id', 'link_id', name='unique_webpage_
    backlink')
                )
metadata.create_all(engine)
```

Let us create a staging table from the dataframe as discussed in Chapter 5 and use the staging table to insert data into these three tables.

```
df.to_sql('staging',con = engine)

insert_into_sources_table = text(
```

```
        "INSERT OR IGNORE INTO sources (source_url) "
        "SELECT webpage_source FROM staging UNION SELECT backlink_source FROM
        staging;"
)
conn.execute(insert_into_sources_table)

insert_into_webpages_table = text(
        "INSERT OR IGNORE INTO webpages (webpage_url, source_id) "
        "SELECT staging.url, sources.source_id "
        "FROM staging, sources "
        "WHERE staging.webpage_source = sources.source_url;"
)

conn.execute(insert_into_webpages_table)

insert_into_webpages_table2 = text(
        "INSERT OR IGNORE INTO webpages (webpage_url, source_id) "
        "SELECT staging.backlink, sources.source_id "
        "FROM staging, sources "
        "WHERE staging.backlink_source = sources.source_url;"
)

conn.execute(insert_into_webpages_table2)

insert_into_backlinks_table = text(
        "INSERT  OR IGNORE INTO backlinks (link_id, webpage_id) "
        "SELECT E.webpage_id, F.webpage_id "
        "FROM webpages AS E,webpages AS F, staging "
        "WHERE E.webpage_url = staging.backlink "
        "AND F.webpage_url = staging.url;"
)

conn.execute(insert_into_backlinks_table)

conn.execute(text("DROP TABLE staging;"))
```

Now that the database is completely loaded up with backlinks data, let us query for web page URLs containing the highest number of backlinks in Listing 6-29.

Listing 6-29. Querying for the most popular backlinks in our database

```
sample_query = '''SELECT
    link_id,webpages.webpage_url as backlink_url, COUNT(link_id)
FROM
    backlinks, webpages

    where
        backlinks.link_id = webpages.webpage_id
GROUP BY
    link_id
HAVING
    COUNT(link_id) > 2
ORDER BY COUNT(link_id) DESC
LIMIT 1000;'''

query = conn.execute(sample_query)
#results_list = conn.execute(text(sample_query)).fetchall()

result_list = query.fetchall()
result_list_keys = query.keys()

df = pd.DataFrame(result_list, columns = result_list_keys)
df.head(10)
```

	link_id	backlink_url	Total_Count
0	53825	https://twitter.com/share	1164
1	53454	https://wordpress.org/	793
2	58166	https://www.blogger.com	664
3	148565	https://automattic.com/cookies	658
4	645301	https://wordpress.com/?ref=footer_blog	612
5	148563	https://gravatar.com/site/signup/	597
6	94565	http://wordpress.org/	421
7	51538	https://akismet.com/privacy/	406
8	86804	http://twitter.com/share	320
9	1121506	https://www.shopify.com?utm_campaign=poweredby...	289

We notice right away that it's dominated by web page URLs from WordPress, Twitter, Blogger, and so on, so nothing unexpected here; however, we could've probably normalized the URLs a bit more and mapped the "http://" and "https://" pages into a single unique URL.

We can get a bigger picture by querying for top domain and host-level backlinks in Listing 6-30. This is also calculated by the Ahrefs dataset.

Listing 6-30. Querying for the most popular domains in our database

```
sample_query = '''SELECT
    sources.source_id, sources.source_url AS Source_Url, COUNT(sources.
source_url) AS Total_Count
FROM
    backlinks, webpages, sources

    where
        backlinks.link_id = webpages.webpage_id
    AND
        webpages.source_id = sources.source_id

GROUP BY
    Source_Url
HAVING
    Total_Count > 2
ORDER BY Total_Count DESC
LIMIT 1000;'''

query = conn.execute(sample_query)

result_list = query.fetchall()
result_list_keys = query.keys()

df = pd.DataFrame(result_list, columns = result_list_keys)
df.head(10)
```

	source_id	Source_Url	Total_Count
0	152418	wordpress.com	122145
1	47205	facebook.com	32291
2	19530	blogger.com	31007
3	142405	twitter.com	29723
4	48099	fc2.com	23276
5	128527	spartantown.net	14649
6	67805	instagram.com	14507
7	56792	google.com	13593
8	155985	youtube.com	11498
9	11662	arpati.blogspot.com	10039

If you are indexing more than a few hundred WAT files into the database, then I would recommend switching to RDS-based PostgreSQL which provides the ability to store a maximum of 64 TB which is more than enough to store backlinks from all the WAT files from multiple monthly crawls.

There are many applications for backlinks databases other than SEO, but by far the most common one is using them to calculate domain-level rankings and domain authority using harmonic centrality or PageRank algorithms. They can also be used to calculate page authority as a measure of relative popularity of a certain web page for a given domain address. The Ahrefs database captures both of these metrics as seen in Figure 1-6 of Chapter 1-6.

Common crawl provides host- and domain-level graphs periodically (`https://commoncrawl.org/2020/06/host-and-domain-level-web-graphs-febmarmay-2020/`) by combining WAT files from multiple monthly crawls. The domain ranking files are about 2 GB compressed, so it will be difficult for many readers to fit them entirely in memory to read them. Hence, we will introduce Amazon Athena in the next chapter which can be used to query the data located in S3 using SQL queries.

Summary

We discussed publicly accessible web crawl datasets from the Common Crawl Foundation available on AWS's registry of open data and used them to create a website similarity, web technology profiling, and backlinks database.

In the next chapter, we will learn about performing web crawl processing on a big data scale by using Amazon Athena and a distributed computing architecture.

## CHAPTER 7

Web Crawl Processing on Big Data Scale

In this chapter, we'll learn about processing web crawl data on a big data scale using distributed computing architecture using Amazon Web Services (AWS).

There are distinct advantages to processing the data where it is stored, so that we do not waste our server time on downloading the data which is rate limiting based on your Internet speed.

We will also learn about Amazon Athena which can be used to query data located in S3 using the SQL language without setting up a server.

The overall goal of this chapter is to get you to a stage where you can efficiently process a large fraction of the common crawl dataset and populate the database we created in Chapter 5.

In the last section of this chapter, we will revisit the sentiment analysis example from Chapter 1 and show you how it can become a commercially relevant dataset for application in alternative financial analysis.

Domain ranking and authority using Amazon Athena

Amazon Athena is a serverless service built on Presto (`https://prestodb.io/`) that lets us query the data located in an S3 bucket by paying (as of the time of publication) $5 per TB of data scanned with a minimum of 10 MB per query.

Athena supports a variety of data formats such as parquet, CSV, txt, and so on, along with compression and partitioning so that you can reduce the amount of data scanned, thereby reducing the overall cost per query.

© Jay M. Patel 2020
J. M. Patel, *Getting Structured Data from the Internet*, https://doi.org/10.1007/978-1-4842-6576-5_7

We will load the domain ranking dataset from common crawl, and once we are comfortable with it, we will set up the common crawl index in Athena.

Data definition language (DDL) type queries are free in Athena; you are only charged for actual queries. The queried results are saved as a CSV file in the S3 bucket configured by you in Athena. We can also directly fetch the data using boto3 and paginate through the results if needed.

There is a learning curve to writing DDL queries in Athena since it supports various file formats all of which require different row delimitation parameters. In this chapter, we will create tables out of three file formats such as txt.gz, CSV, and parquet so that you get a better idea on the DDL syntax.

Let us create a new database in Athena using the boto3 package in Listing 7-1. You will need to specify a query folder on S3 where Athena will store the query results. Before you start using Athena, make sure you log in to your AWS account using root credentials and assign an AWS-managed policy called "AmazonAthenaFullAccess" for the IAM user that is being used with boto3.

Listing 7-1. Creating a new database in Athena

```
import boto3
import numpy as np
import pandas as pd

def run_query(query, database, s3_output):
    client = boto3.client('athena', region_name='us-east-1')
    response = client.start_query_execution(
        QueryString=query,
        QueryExecutionContext={
            'Database': database
            },
        ResultConfiguration={
            'OutputLocation': s3_output,
            }
        )
    print('Execution ID: ' + response['QueryExecutionId'])
    return response['QueryExecutionId']

query = '''Create database domainranks2'''
```

```
database = 'domainranks2'
s3_output = 's3://athena-us-east-1-testing/query-folder2/'
execution_id = run_query(query, database, s3_output)

query = '''Create database domainranks2'''

database = 'domainranks2'
s3_output = 's3://athena-us-east-1-testing/query-folder2/'
execution_id = run_query(query, database, s3_output)
```

Create a folder on S3 called domain_ranks and upload the domain ranks file from the common crawls S3 bucket (https://commoncrawl.s3.amazonaws.com/projects/ hyperlinkgraph/cc-main-2019-20-nov-dec-jan/domain/cc-main-2019-20-nov-dec- jan-domain-ranks.txt.gz).

We discussed in Chapter 1 about how search engines determine domain authority based on algorithms such as PageRank and harmonic centrality which take into account the number and quality of backlinks or inlinks. The common crawl project computes PageRank and harmonic centrality metrics for all the domain addresses it crawls by combining three monthly crawls and makes the data publicly available sorted by the rank of the relative importance. This is computed by link inversion from WAT records similar to how we did it in Chapter 6.

You can manually download this and upload it to your bucket via Cyberduck or programmatically by using boto3. We got this domain ranks file and column names from the common crawl blog post (https://commoncrawl.org/2020/02/host-and-domain- level-web-graphs-novdecjan-2019-2020/); they publish new domain ranks about four times a year so that you can get the latest file path or column changes by checking the blog posts.

Once the file is uploaded to the domain_ranks folder, let's create a new table as shown in Listing 7-2 that reads data from this file.

Listing 7-2. Creating a new table

```
query = '''CREATE EXTERNAL TABLE IF NOT EXISTS domainranks2.domain_ranks (
`#harmonicc_pos` bigint,
`#harmonicc_val` double,
`#pr_pos` bigint,
`#pr_val` double,
```

```
`#host_rev` string,
`#n_hosts` int)
ROW FORMAT SERDE 'org.apache.hadoop.hive.serde2.lazy.LazySimpleSerDe'
WITH SERDEPROPERTIES (
  'serialization.format' = '    ',
  'field.delim' = '    ',
  'collection.delim' = '#',
  'mapkey.delim' = '#'
) LOCATION 's3://athena-us-east-1-testing/domain_ranks/'
TBLPROPERTIES ('has_encrypted_data'='false')'''

database = 'domainranks2'
s3_output = 's3://athena-us-east-1-testing/query-folder2/'
execution_id = run_query(query, database, s3_output)
#Output
Execution ID: 5d41677a-3592-42e6-b250-c995987e5770
```

Let's test out this new database by querying for ranks of theguardian.com in Listing 7-3. Note that the host_rev is based on inverted URL format, so we will have to query for "com. theguardian".

Listing 7-3. Querying for ranks for theguardian.com

```
query = '''SELECT * FROM domainranks2.domain_ranks where domain_
ranks."#host_rev" = 'com.theguardian';'''
database = 'domainranks2'
s3_output = 's3://athena-us-east-1-testing/query-folder2/'
execution_id = run_query(query, database, s3_output)
# Output
Execution ID: caa1997d-3625-4598-920b-6f3058ecc742
```

We get query results by sending in a request with execution_id; these queries can take 1–5 minutes depending on the size of the queried table, so it's a good idea to include multiple retries when fetching results from Athena as shown in Listing 7-4. We'll also have to do some data wrangling to convert the raw data from an Athena query into a usable form which is ready to be loaded in a pandas dataframe. We should create a boto3 client in the same region as the S3 bucket which in the case of common crawl datasets is us-east-1.

Listing 7-4. Fetching query results from Athena

```python
def get_raw_response(execution_id):

    client = boto3.client('athena', region_name='us-east-1')
    response = client.get_query_results(
        QueryExecutionId=execution_id,
        MaxResults=123
    )
    return response

def results_to_df(results):

    columns = [
        col['Label'] for col in results['ResultSet']['ResultSetMetadata']
        ['ColumnInfo']]

    listed_results = []
    for res in results['ResultSet']['Rows'][1:]:
        values = []
        for field in res['Data']:
            try:
                values.append(list(field.values())[0])
            except:
                values.append(list(' '))

        listed_results.append(
            dict(zip(columns, values))
        )

    return listed_results

import time
for i in range(15):
    time.sleep(10)

    try:
        return_json = get_raw_response(execution_id)
        t = results_to_df(return_json)
```

```
        df_2 = pd.DataFrame(t)
        print("query successful")
        break
    except Exception as e:

        print(e)

        pass
df_2.head()
# Output
query successful
```

#harmonicc_pos	#harmonicc_val	#host_rev	#n_hosts	#pr_pos	#pr_val
0 92	2.043555E7	com. theguardian	127	171	1.4701598631846095E-4

A harmonic centrality ranking of 92 for theguardian.com is pretty similar with the 52 obtained with the majestic million dataset (https://majestic.com/reports/majestic-million?domain=theguardian.com&DefaultSearchText=example1.com%2Cexample2.com) and 111 with Alexa (http://data.alexa.com/data?cli=10&url=theguardian.com) so that tells us that the common crawl graph database is comparable with commercial providers.

There is some debate in literature about whether harmonic centrality–based rankings (#harmonicc_pos) are better than PageRank rankings (#pr_pos) in capturing domain authority; I think it doesn't matter much for us as long as we stick to one of the metrics for all our analysis. Similarly, the raw values of harmonic centrality (#harmonicc_val) and page ranks (#pr_val) are useful for calculating a normalized score (0–100) for domain authority if you are going to combine this with harmonic and PageRank values for your own crawled data.

Batch querying for domain ranking and authority

In this section, we will fetch domain-level rankings on hundreds or even thousands of domains in a single Athena query. We get charged in Athena on the basis of scanned data, so you get charged the same amount regardless of how much data gets returned, and hence it's best to query for bulk results instead of individual domain ranks like we did in Listing 7-4.

It supports SQL joins so that you can request domain rankings in bulk by first creating a separate table and using it in your SQL query.

I have created a CSV file with just two columns, a domain and an inverted domain address as shown in Listing 7-5; we will load this up on our Athena database as a new table.

Listing 7-5. Inverted URL dataframe

```
df = pd.read_csv("inverted_urls_list.csv")
df.head()

# Output
```

	Urls	inverted_urls
0	facebook.com	com.facebook
1	google.com	com.google
2	youtube.com	com.youtube
3	twitter.com	com.twitter
4	instagram.com	com.instagram

You should upload this to an S3 folder and mention its location in the query as shown in Listing 7-6.

Listing 7-6. Creating an Athena table with inverted URLs

```
query = '''CREATE EXTERNAL TABLE IF NOT EXISTS urltest (
 `urls` STRING,
 `inverted_urls` STRING)
ROW FORMAT SERDE 'org.apache.hadoop.hive.serde2.OpenCSVSerde'
```

```
WITH SERDEPROPERTIES (
  'escapeChar'='\\\\',
  'separatorChar'=','
  ) LOCATION 's3://athena-us-east-1-testing/sample-folder/'
TBLPROPERTIES ('has_encrypted_data'='false', 'skip.header.line.count' = '1');'''

database = 'domainranks2'
s3_output = 's3://athena-us-east-1-testing/query-folder2/'
execution_id = run_query(query, database, s3_output)
```

We should test whether Athena is able to read the CSV file correctly by calling a sample query shown in Listing 7-7.

Listing 7-7. Testing the new Athena table

```
query = '''select * from domainranks2.urltest'''
database = 'domainranks2'
s3_output = 's3://athena-us-east-1-testing/query-folder2/'
execution_id = run_query(query, database, s3_output)

return_json = get_raw_response(execution_id)
t = results_to_df(return_json)
df_2 = pd.DataFrame(t)
df_2.head()
# Output
```

	inverted_urls	urls
0	com.facebook	facebook.com
1	com.google	google.com
2	com.youtube	youtube.com
3	com.twitter	twitter.com
4	com.instagram	instagram.com

Listing 7-8 shows a query for fetching harmonic centrality and PageRank rankings and scores by using a join across urltest and domainranks tables.

Listing 7-8. Fetching harmonic centrality and PageRank rankings

```
query = '''select urltest."inverted_urls",urltest."urls", domain_
ranks."#harmonicc_pos", domain_ranks."#harmonicc_val", domain_ranks.
"#n_hosts", domain_ranks."#pr_pos", domain_ranks."#pr_val"
FROM urltest, domain_ranks
WHERE domain_ranks."#host_rev" = urltest."inverted_urls"'''
database = 'domainranks2'
s3_output = 's3://athena-us-east-1-testing/query-folder2/'
execution_id = run_query(query, database, s3_output)
import time
for i in range(15):
    time.sleep(10)

    try:
        return_json = get_raw_response(execution_id)
        t = results_to_df(return_json)
        df_2 = pd.DataFrame(t)
        print("query successful")
        break
    except Exception as e:

        print(e)

        pass
df_2.head()
```

`#Output`

	#harmonicc_ pos	#harmonicc_ val	#n_ hosts	#pr_ pos	#pr_val	inverted_ urls	urls
0	24	2.153328E7	100	31	0.0011040749131266557	com.vimeo	vimeo. com
1	80	2.0488868E7	5116	373	6.71641116389417E-5	com.msn	msn. com
2	67	2.0609614E7	464	154	1.692129080835582E-4	com.bing	bing. com
3	236490	1.6137525E7	17	116	2.57561600845181E-4	com.cpanel	cpanel. com
4	146	2.0222314E7	123	339	7.442682626049672E-5	com.time	time. com

Processing parquet files for a common crawl index

We have seen the usefulness of the common crawl index API in fetching raw web pages from specific domain addresses. However, it is still limited due to the following reasons:

- We need different API endpoints for getting captures from each month's crawls. For example, if you want to fetch all the pages captured by common crawl for theguardian.com in the past two years, you will have to make 24 separate API calls not including pagination for each of them.

- Certain time-consuming queries such as fetching all the pages with .com tld such as this "*.com" will be timed out, and the API server will return back a 502 error before returning any results. Even tlds with fewer pages such as *.ai have been known to give errors too.

- You cannot query for multiple domain addresses in a single API query. It's an extremely common use case to get all the common crawl captures from top 10,000 news websites or, in an even broader case, get all captures from top 1 million Alexa or majestic domains for performing various kinds of natural language processing tasks.

All of the preceding use cases are possible if you either process all of the data from the common crawl index on your servers using Apache Spark or set up an Amazon Athena database for doing it.

In this section, we will write a pretty complicated DDL query to create a table in our Athena database using data from the common crawl index in parquet format as shown in Listing 7-9.

Listing 7-9. Setting up a CC index Athena table

```
query = '''CREATE EXTERNAL TABLE IF NOT EXISTS ccindex (
  `url_surtkey`                  STRING,
  `url`                          STRING,
  `url_host_name`                STRING,
  `url_host_tld`                 STRING,
  `url_host_2nd_last_part`       STRING,
  `url_host_3rd_last_part`       STRING,
  `url_host_4th_last_part`       STRING,
  `url_host_5th_last_part`       STRING,
  `url_host_registry_suffix`     STRING,
  `url_host_registered_domain`   STRING,
  `url_host_private_suffix`      STRING,
  `url_host_private_domain`      STRING,
  `url_protocol`                 STRING,
  `url_port`                     INT,
  `url_path`                     STRING,
  `url_query`                    STRING,
  `fetch_time`                   TIMESTAMP,
  `fetch_status`                 SMALLINT,
  `fetch_redirect`               STRING,
  `content_digest`               STRING,
  `content_mime_type`            STRING,
  `content_mime_detected`        STRING,
  `content_charset`              STRING,
  `content_languages`            STRING,
  `content_truncated`            STRING,
  `warc_filename`                STRING,
```

```
    `warc_record_offset`                INT,
    `warc_record_length`                INT,
    `warc_segment`                      STRING)
PARTITIONED BY (
    `crawl`                             STRING,
    `subset`                            STRING)
STORED AS parquet
LOCATION 's3://commoncrawl/cc-index/table/cc-main/warc/';'''

database = 'domainranks2'
s3_output = 's3://athena-us-east-1-testing/query-folder2/'
execution_id = run_query(query, database, s3_output)
```

We should always run the repair query once tables containing partitions such as the one in Listing 7-10 are created so that they use the most updated data. Just like the DDL, these queries are free to run.

Listing 7-10. Table repair query

```
query = '''MSCK REPAIR TABLE ccindex;'''

database = 'domainranks2'
s3_output = 's3://athena-us-east-1-testing/query-folder2/'
execution_id = run_query(query, database, s3_output)
```

Lastly, let's run a sample query on the common crawl index table as shown in Listing 7-11.

Listing 7-11. Sample query on cc-index

```
query = '''SELECT url_surtkey, url, warc_filename, warc_record_offset,
warc_record_length,
content_mime_detected, fetch_status, content_languages
FROM domainranks2.ccindex
WHERE crawl = 'CC-MAIN-2020-24'
AND url LIKE 'http%://www.cnn.com/'
AND subset = 'warc'
AND url_host_registered_domain = 'cnn.com'
LIMIT 5;'''

database = 'domainranks2'
```

```
s3_output = 's3://athena-us-east-1-testing/query-folder2/'
execution_id = run_query(query, database, s3_output)

import time
for i in range(15):
    time.sleep(10)

    try:
        return_json = get_raw_response(execution_id)
        t = results_to_df(return_json)
        df_2 = pd.DataFrame(t)
        print("query successful")
        break
    except Exception as e:

        print(e)

        pass
df_2.head()
#Output
```

content_ languages	content_ mime_ detected	fetch_ status	url	url_ surtkey	warc_filename	warc_ record_ length	warc_ record_ offset
0 eng	text/html	200	https:// www.cnn. com/	com, cnn)/	crawl-data/CC-MAIN-2020-24/ segments/ 1590347387...	143892	693352534
1 eng	text/html	200	https:// www.cnn. com/	com, cnn)/	crawl-data/CC-MAIN-2020-24/ segments/ 1590347387...	144002	676690049

(continued)

content_languages	content_mime_detected	fetch_status	url	url_surtkey	warc_filename	warc_record_length	warc_record_offset
2 eng	text/html	200	https://www.cnn.com/	com, cnn)/	crawl-data/CC-MAIN-2020-24/segments/1590347388...	144422	697305859
3 eng	text/html	200	https://www.cnn.com/	com, cnn)/	crawl-data/CC-MAIN-2020-24/segments/1590347388...	144466	647676538
4 eng	text/html	200	https://www.cnn.com/	com, cnn)/	crawl-data/CC-MAIN-2020-24/segments/1590347389...	144118	680989734

I will leave it as an exercise for you to query for file paths and offsets for hundreds of thousands of domains if necessary by creating an additional table containing domain names and using SQL join over the ccindex table to get all the captures in the common crawl corpus.

Parsing web pages at scale

We have learned about parsing web pages using Beautifulsoup, lxml, XPaths, and Selenium in Chapter 2, and any of those methods can extract information from web pages with ease.

However, it requires you to know about the structure of the web page, and while that is easy enough when we are only parsing content from a handful of web domains, it becomes almost impossible to do it for web pages from millions of domains. There are some proof-of-concept approaches using machine learning to "understand" and parse the structure of a raw HTML, but they are not ready for production use yet. Partial tree alignment–based algorithms such as DEPTA also work for limited use cases, but using them on broad web crawls is very expensive computationally.

There are two ways to deal with this; the first is relying on information located in the meta tags of the web page, just like a search engine would do. We saw meta tags in usage in Figure 1-4 of Chapter 1; and just to refresh your memory, these are optional tags inserted by webmasters to help search engines find relevant content, and while they are invisible to someone actually visiting the page, the content from meta tags is directly used to populate the title and description snippet of search engine results. Due to this, filling out information in meta tags correctly is one of the essential measures of on-site search engine optimization (SEO).

These SEO tags are generally filled manually, so in a way, it is a manually curated summary of the information in the web page, and that is good enough for lots of NLP use cases.

Microdata, microformat, JSON-LD, and RDFa

We can rely on structured data such as microdatamicroformat, JSON-LD, and RDFa on a web page related to the semantic web initiative (`www.w3.org/2001/sw/Activity`) and schema.org, which aims to make the web pages more machine readable.

All of these meta tags or structured data are optional so we will not see all of them on a given web page, but chances are that there will be at least a title tag and some meta tags on a page, and if we are lucky, we will also get one of the structured data such as JSON-LD on a page. The Data and Web Science Research Group at the University of Mannheim regularly extracts structured data out of the common crawl's monthly web crawl (`http://webdatacommons.org/structureddata/#results-2019-1`), and their data shows that about 1 TB of structured data is present in each month's crawl. You are encouraged to check out their detailed analysis (`http://webdatacommons.org/structureddata/2019-12/stats/stats.html`) and download the extracted structured data from the S3 bucket if needed.

Let us explore these structured data in greater detail using a Python package called extruct (pip install extruct). Instead of explaining each of these data formats, we will just illustrate an example by fetching a news article from theguardian.com and showing outputs from three structured data formats.

Listing 7-12 shows parsed microdata, and it is directly apparent that it contains all the important elements from the news article such as the date and published time, author names, title, full text, and image URLs. Along with each data, it also contains reference to the entity type from schema.org so that entity disambiguation can be performed easily.

Listing 7-12. Microdata example

```
import requests

url = 'https://www.theguardian.com/business/2020/feb/10/waitrose-to-launch-
charm-offensive-as-ocado-switches-to-ms'
my_headers = {
'User-Agent': 'Mozilla/5.0 (Windows NT 10.0; Win64; x64) AppleWebKit/537.36
' + ' (KHTML, like Gecko) Chrome/61.0.3163.100Safari/537.36'
}

r = requests.get(url, headers = my_headers)
html_response = r.text

data = extruct.extract(r.text, syntaxes = ['microdata'])
data_keys = data["microdata"][1]["properties"].keys()
for key in data_keys:
    print("*"*10)
    print(key)
    print(data["microdata"][1]["properties"][key])
# Output

**********
mainEntityOfPage
https://www.theguardian.com/business/2020/feb/10/waitrose-to-launch-charm-
offensive-as-ocado-switches-to-ms
**********
publisher
{'type': 'https://schema.org/Organization', 'properties': {'name': 'The
Guardian', 'logo': {'type': 'https://schema.org/ImageObject', 'properties':
{'url': 'https://uploads.guim.co.uk/2018/01/31/TheGuardian_AMP.png',
'width': '190', 'height': '60'}}}}
**********
headline
Waitrose to launch charm offensive as Ocado switches to M&S
**********
description
```

Supermarket will launch thousands of new and revamped products aiming to retain online customers
****
author
{'type': 'http://schema.org/Person', 'properties': {'sameAs': 'https://www.theguardian.com/profile/zoewood', 'name': 'Zoe Wood'}}
****
datePublished
2020-02-10T06:00:27+0000
****
dateModified
['2020-02-12T13:42:07+0000', '2020-02-12T13:42:07+0000']
****
associatedMedia
{'type': 'http://schema.org/ImageObject', 'properties': {'representativeOfPage': 'true', 'url': 'https://i.guim.co.uk/img/media/65d537a07a3493f18eef074ac0910e6c768d5f2c/0_58_3500_2100/master/3500.jpg?width=700&quality=85&auto=format&fit=max&s=8c4fdd2153ed244918a8a293c24d4f6e', 'width': '3500', 'height': '2100', 'contentUrl': 'https://i.guim.co.uk/img/media/65d537a07a3493f18eef074ac0910e6c768d5f2c/0_58_3500_2100/master/3500.jpg?width=300&quality=85&auto=format&fit=max&s=4c0443760fc652dad651f55c4bfce7cc', 'description': 'Analysts suggest the end of the Ocado deal may have badly affect Waitrose's owner, the John Lewis Partnership. Photograph: Andrew Matthews/PA'}}
****
image
{'type': 'http://schema.org/ImageObject', 'properties': {'representativeOfPage': 'true', 'url': 'https://i.guim.co.uk/img/media/65d537a07a3493f18eef074ac0910e6c768d5f2c/0_58_3500_2100/master/3500.jpg?width=700&quality=85&auto=format&fit=max&s=8c4fdd2153ed244918a8a293c24d4f6e', 'width': '3500', 'height': '2100', 'contentUrl': 'https://i.guim.co.uk/img/media/65d537a07a3493f18eef074ac0910e6c768d5f2c/0_58_3500_2100/master/3500.jpg?width=300&quality=85&auto=format&fit=max&s=4c0443760fc652dad651f55c4bfce7cc', 'description': 'Analysts suggest the end of the Ocado deal may

341

have badly affect Waitrose's owner, the John Lewis Partnership. Photograph: Andrew Matthews/PA'}}
****
articleBody
Waitrose is to launch thousands of new and revamped products in the coming months as the battle for the hearts and minds of Ocado shoppers moves up a gear.

The supermarket's deal with the online grocer will finish at the end of August, when it will be replaced by Marks & Spencer. The switchover is high risk for all the brands involved: Ocado risks losing loyal Waitrose shoppers while the supermarket, which is part of the John Lewis Partnership, will have to persuade shoppers to use its own website instead.

. . . (text truncated)

Listings 7-13 and 7-14 show other popular structured data formats known as JSON-LD and opengraph, respectively, which contain less information than microdata format but still pretty useful where microdata is not present on a web page.

Listing 7-13. JSON-LD example

```
data = extruct.extract(html_response, syntaxes = ['json-ld'])
print(data)
# Output
{'json-ld': [{'@context': 'http://schema.org',
   '@type': 'Organization',
   'logo': {'@type': 'ImageObject',
    'height': 60,
    'url': 'https://uploads.guim.co.uk/2018/01/31/TheGuardian_AMP.png',
    'width': 190},
   'name': 'The Guardian',
   'sameAs': ['https://www.facebook.com/theguardian',
    'https://twitter.com/guardian',
    'https://www.youtube.com/user/TheGuardian'],
   'url': 'http://www.theguardian.com/'},
  {'@context': 'http://schema.org',
```

```
'@id': 'https://www.theguardian.com/business/2020/feb/10/waitrose-to-
launch-charm-offensive-as-ocado-switches-to-ms',
'@type': 'WebPage',
'potentialAction': {'@type': 'ViewAction',
 'target': 'android-app://com.guardian/https/www.theguardian.com/
business/2020/feb/10/waitrose-to-launch-charm-offensive-as-ocado-
switches-to-ms'}}]}
```

Listing 7-14. opengraph example

```
data = extruct.extract(html_response, syntaxes = ['opengraph'])
print(data)
#Output
{'opengraph': [{'namespace': {'article': 'http://ogp.me/ns/article#',
   'og': 'http://ogp.me/ns#'},
  'properties': [('og:url',
    'http://www.theguardian.com/business/2020/feb/10/waitrose-to-launch-
    charm-offensive-as-ocado-switches-to-ms'),
   ('article:author', 'https://www.theguardian.com/profile/zoewood'),
   ('og:image:height', '720'),
   ('og:description',
    'Supermarket will launch thousands of new and revamped products aiming
    to retain online customers'),
   ('og:image:width', '1200'),
   ('og:image',
    'https://i.guim.co.uk/img/media/65d537a07a3493f18eef074ac0910e6c7
    68d5f2c/0_58_3500_2100/master/3500.jpg?width=1200&height=630&qual
    ity=85&auto=format&fit=crop&overlay-align=bottom%2Cleft&overlay-
    width=100p&overlay-base64=L2ltZy9zdGF0aWMvb3ZlcmxheXMvdGctZGVmYXVsdC5w
    bmc&enable=upscale&s=9719e60266c3af3c231324b6969a0c84'),
   ('article:publisher', 'https://www.facebook.com/theguardian'),
   ('og:type', 'article'),
   ('article:section', 'Business'),
   ('article:published_time', '2020-02-10T06:00:27.000Z'),
   ('og:title',
    'Waitrose to launch charm offensive as Ocado switches to M&S'),
```

```
('article:tag',
 'Waitrose,Ocado,Business,UK news,Retail industry,Online shopping,
  Marks & Spencer,John Lewis,Supermarkets,Money'),
('og:site_name', 'the Guardian'),
('article:modified_time', '2020-02-12T13:42:07.000Z')]}]}
```

We will exclude showing other types of parsed structured data formats such as microformat and RDFa, but the extruct package handles them as well.

Parsing news articles using newspaper3k

Structured data formats discussed in the earlier section are a great way to extract information from web pages when they are available. Despite the obvious SEO-related advantages, there are still a sizable number of websites which do not contain any structured data on their web pages except perhaps only a title tag.

We can still extract information from these pages by relying on some generalized rules and a brute-force approach on extracting required data on a web page.

For example, the common CSS class names for authors are "name", "itemprop", "class", and "id"; similarly, the common attributes for author class names are "author," "byline," and so on.

Similarly, there are a few ways web pages expose the dates of the articles; the simplest is by encoding them in the URL itself, and that can be extracted by a simple function shown in Listing 7-15. If the date cannot be extracted this way, then you can try a couple of other approaches.

Listing 7-15. Extracting dates from a URL

```
import re
from dateutil.parser import parse as date_parser

def extract_dates(url):

    def parse_date_str(date_str):
        if date_str:
            try:
                return_value = date_parser(date_str)
                if pd.isnull(return_value) is True:
```

```
                    return 'None'
            else:
                return return_value
        except (ValueError, OverflowError, AttributeError, TypeError):

            return 'None'

    _STRICT_DATE_REGEX_PREFIX = r'(?<=\W)'
    DATE_REGEX = r'([\./\-_]{0,1}(19|20)\d{2})[\./\-_]{0,1}(([0-3]{0,1}
[0-9][\./\-_])|(\w{3,5}[\./\-_]))([0-3]{0,1}[0-9][\./\-]{0,1})?'
    STRICT_DATE_REGEX = _STRICT_DATE_REGEX_PREFIX + DATE_REGEX

    date_match = re.search(STRICT_DATE_REGEX, url)

    if date_match is not None:

        return parse_date_str(date_match.group(0))
    else:
        return 'None'
url = 'https://www.theguardian.com/business/2020/feb/10/waitrose-to-launch-
charm-offensive-as-ocado-switches-to-ms'

print(extract_dates(url))
# Output
2020-02-10 00:00:00
```

Newspaper3k is a very handy package that implements lots of these approaches using the lxml library for parsing information from web pages mainly from news websites. It can be installed by simply running "pip install newspaper3k". Listing 7-16 parses the same Guardian article using it. In the next section, we will revisit sentiment analysis and use newspaper3k to parse web pages on a distributed big data scale.

Listing 7-16. Newspaper3k parsing example

```
import json
from newspaper import Article
import numpy as np
import pandas as pd
import requests
```

```python
url = 'https://www.theguardian.com/business/2020/feb/10/waitrose-to-launch-
charm-offensive-as-ocado-switches-to-ms'

my_headers = {
'User-Agent': 'Mozilla/5.0 (Windows NT 10.0; Win64; x64) AppleWebKit/537.36
' + ' (KHTML, like Gecko) Chrome/61.0.3163.100Safari/537.36'
}

r = requests.get(url, headers = my_headers)

html_response = r.text

def newspaper_parse(html):

    article = Article('')
    article.set_html(html)
    article.download()

    article_title = None
    json_authors = None
    article_text = None
    article_publish_date = None

    try:

        article.parse()

        json_authors = json.dumps(article.authors)

        article_title = article.title
        article_text = article.text
        article_publish_date = article.publish_date

    except:

        pass

    return article_title, json_authors, article_text, article_publish_date

article_title, json_authors, article_text, article_publish_date =
newspaper_parse(html_response)
print(article_title)
```

```
print("*"*10)
print(article_publish_date)
print("*"*10)
print(json_authors)
print("*"*10)
print(article_text)
# Output
Waitrose to launch charm offensive as Ocado switches to M&S
**********

2020-02-10 06:00:27+00:00
**********

["Zoe Wood"]
**********
```

Waitrose is to launch thousands of new and revamped products in the coming months as the battle for the hearts and minds of Ocado shoppers moves up a gear.

. . .

(Output truncated)

Revisiting sentiment analysis

In Figures 1-10 and 1-11 of Chapter 1, we attempted to use sentiment analysis on user posts by searching for a company name (Exxon) in a particular subreddit called investing and tried to see if there was any way it can serve as a trading signal by comparing to its stock price. There were lots of potential issues preventing it from being effective; let's go through all the steps necessary to make our sentiment analysis more robust and similar to commercial data providers.

- We had too few data points in Chapter 1 to feed to a sentiment analysis model even if there weren't any other issues, and this can now be easily fixed thanks to the access to the common crawl dataset which can be used as a one-stop source to extract news stories from multiple outlets at once.

- A simple keyword-based searching of the company name itself is not enough to uncover all the potential news stories relevant to a company; it's much better to compare a set of tokens such as company's major brands, subsidiaries, and so on against extracted named entities from news stories for better relevancy. This step basically links a particular news story to one or many stock ticker symbols. For example, InfoTrie (`www.quandl.com/databases/NS1/documentation`) includes coverage for over 49,000 financial securities across the world, allowing the customers to query news sentiments data just by specifying the stock ticker symbol. Many providers also use negative lists to exclude certain news articles to ensure proper disambiguation and decreasing overall noise.

- Once you start extracting text from multiple domain addresses, you have to take weighted averages for sentiments according to their relative audience reach. For example, a mildly negative sentiment of a document hosted on .gov (official US government page) is way more adverse than a strongly negative document on a .vu country code TLD which represents the country code for Vanuatu, a tiny nation in the South Pacific Ocean. If a media outlet is online only, then the relative audience outreach can be inferred by checking their domain authority deduced from harmonic or PageRank-based domain rankings. However, there are other forms of media outlets like television stations, radio, newspaper, and so on for which their website is a secondary mode of news delivery, and hence domain authority may not capture their real audience outreach. In these cases, it's essential to approximate the audience outreach by using surrogate measures like aggregate social media following for a particular media outlet or better still using industry audience survey results which are easy to find for traditional outlets like print and television media.

- It's a good idea to correct any media bias in news articles published at a domain address toward a specific company or sector. For example, certain polarizing industries such as the fossil fuel industry, cigarettes, and so on generate a lot of sector-specific negative bias in media outlets, and this needs to be factored in when trying to capture sentiment scores from articles published at such outlets. The main motivation behind it is that a strongly negative sentiments story coming from a negatively biased media outlet about an industry will not serve as a strong enough signal for financial markets as say the same type of news story with the same sentiments score coming from a relatively unbiased newswire service like Reuters or Associated Press.

- We should establish an approximate geographical coverage of each news outlet through consistently tracking geographical location-based entities from each news story. A strong sentiment scoring from highly regional outlets about a company should be flagged for manual checking to ensure it's not a case of strongly negative uncertain events like chemical spills, explosions, and so on.

- Author names should be extracted and an influence score calculated. One of the most robust ways to calculate this is by considering the number and quality of backlinks of authors' past articles. We can easily do that by keeping an updated backlinks database like we saw in Chapter 6. You can also take into account the number of followers on social media or some other indirect metrics.

- Author name disambiguation is a potential issue when we are trying to distinguish authors with the same or near duplicate names. There are lots of methods out there, but the easiest one is doing it on the basis of email addresses and social media handles.

- An author-level bias has to be factored in too so that a sentiment analysis model is robust enough not to be fallen off the tracks by a few articles from influential individuals. For example, this happens when editorials or other articles written by celebrities are talking negatively about a company.

There are many commercially available sentiment data providers in the market right now, and they perform some or most of the steps listed earlier using opaque algorithms since it's their "secret sauce"; but all of them extensively rely on web crawling to get the underlying data.

I frequently see newbie developers and data scientists mistakenly think training a good sentiment classification model is the chief performance hurdle, and they expect it to be the area where companies would be doing the most research.

However, the truth is that the real research is happening to improve author-level disambiguation, calculating the author influence score using an extensive backlinks database and developing a good named entity recognition (NER) model to increase the relevance of text on which the sentiments are calculated.

Another challenge is modifying raw outputs of text sentiments to make them suitable for financial markets lingo. For example, generally speaking, a headline like "some unnamed security yield climbs" will be considered as a positive sentiment; however, when the unnamed security is actually bonds, their yield going up actually means that price is falling. In these cases, we will have to reverse the sentiments by checking the subject of the sentence via a combination of rule-based parsing combined with a dependency parser.

You would need to provide sentiment-level data on a near real-time scale to your client if they are going to use it as a trading signal; however, along with that, you will also need lots of historical sentiment data so that the client can backtest with stock pricing data and see its effectiveness.

We will discuss ways to get media outlet and journalist data in the next section, and later we will talk about distributed computing that will allow you to process data on TB scale, and you can use the same architecture to run web crawler scripts from Listing 2-17 from Chapter 2 to run in parallel on a large scale.

Scraping media outlets and journalist data

We mentioned that you would want to regularly scrape from thousands of media outlets, and based on running text classifiers and NERs, you would want to save information related to geographical coverage, topics (called as "beats" in journalism lingo), and so on, which the outlet covers on a regular basis.

Lots of these require some manual curation, at least initially to ensure the high quality of sources being indexed as well as to determine the relevancy of a source for our business applications. One of the things you can do is publicly scrape a public relations database called muckrack.com to initially populate your database with journalist names, social media handles, website addresses, and so on. All of these will be very useful in the disambiguation between different author names.

Unfortunately, muckrack.com explicitly prohibits common crawl bot (CCBot) from scraping any content of the website from their robots.txt; hence, we have to scrape this on our own in Listing 7-17 using their sitemaps.xml.

Listing 7-17. Scraping from Muck Rack's sitemap

```
import numpy as np
import pandas as pd
import requests

url = 'https://muckrack.com/sitemap.xml'
my_headers = {
'User-Agent': 'Mozilla/5.0 (Windows NT 10.0; Win64; x64) AppleWebKit/537.36
' + ' (KHTML, like Gecko) Chrome/61.0.3163.100Safari/537.36'
}

r = requests.get(url=url, headers = my_headers)
# Listing 7-x: scraping from muckrack's sitemap (cont.)

sitemaps = soup.find_all('sitemap')
for sitemap in sitemaps:
    print(sitemap.find('loc').get_text())
    try:
        print(sitemap.find('lastmod').get_text())
        print(sitemap.find('changefreq').get_text())
    except:
        pass
#Output
https://muckrack.com/sitemaps/sitemap-pages-1.xml
https://muckrack.com/sitemaps/sitemap-mrdaily-1.xml
2020-08-18T12:01:04-04:00
https://muckrack.com/sitemaps/sitemap-mrdaily-2.xml
2016-09-01T17:01:26-04:00
https://muckrack.com/sitemaps/sitemap-mrdaily-3.xml
2014-02-19T10:28:15-05:00
https://muckrack.com/sitemaps/sitemap-blog-1.xml
2020-08-18T06:00:00-04:00
https://muckrack.com/sitemaps/sitemap-media_outlets-1.xml
```

```
2020-08-18T08:57:12-04:00
https://muckrack.com/sitemaps/sitemap-media_outlets-2.xml
2020-08-18T17:54:52-04:00
https://muckrack.com/sitemaps/sitemap-media_outlets-3.xml
2020-08-18T20:35:23-04:00
https://muckrack.com/sitemaps/sitemap-media_outlets-4.xml
2020-08-18T14:29:40-04:00
. . . (output truncated)
https://muckrack.com/sitemaps/sitemap-person-65.xml
2020-08-19T00:56:48-04:00
https://muckrack.com/sitemaps/sitemap-person-66.xml
2020-08-19T00:44:04-04:00
https://muckrack.com/sitemaps/sitemap-person-67.xml
2020-08-18T23:20:22-04:00
```

Muck Rack has split its sitemap into different subject areas, with about 67 pages for sitemaps containing journalist profiles and 11 pages for media outlet profiles. They also have sitemaps for other web pages such as blog posts and so on, but let's ignore them for now. Listing 7-18 loads them up in separate lists.

Listing 7-18. Loading person and media outlet sitemaps into separate lists

```python
from bs4 import BeautifulSoup
soup = BeautifulSoup(r.text, 'xml')
sitemap_other = []
sitemap_media = []
sitemap_persons = []

sitemaps = soup.find_all('loc')
for sitemap in sitemaps:
    sitemap = sitemap.get_text()
    if 'media' in sitemap:
        sitemap_media.append(sitemap)
    elif 'person' in sitemap:
        sitemap_persons.append(sitemap)
    else:
```

```
        sitemap_other.append(sitemap)
print(len(sitemap_media))
print(len(sitemap_persons))
# Output
12
67
```

Listing 7-19 fetches the Muck Rack profile links and last modified date from the media outlet list and saves them as a CSV for later usage. There are about 56,300 media outlets with profiles on Muck Rack which are good enough for implementing any sentiment analysis model for financial markets. We did a similar exercise for extracting journalist profile links, and we got about 231,732 journalist profiles via sitemap. If you are scraping Muck Rack from your local machine, then be very careful about setting a long delay between requests, or you run a real risk of being blocked. If you want to deploy a distributed crawler on muckrack.com, then I highly recommend going through Chapter 8 first and implementing some measures such as IP rotation to ensure a successful crawling.

Listing 7-19. Fetching Muck Rack profiles URL from the sitemap list

```
import time
temp_list = []
for sitemap_media_url in sitemap_media:
    time.sleep(5)
    r = requests.get(url = sitemap_media_url, headers = my_headers)
    soup = BeautifulSoup(r.text, 'xml')
    sitemaps = soup.find_all('url')
    for sitemap in sitemaps:

        temp_dict = {}
        temp_dict['url'] = sitemap.find('loc').get_text()

        try:
            last_modified = sitemap.find('lastmod').get_text()

        except:
            last_modified = ''
```

```
        temp_dict["last_modified"] = last_modified
        temp_list.append(temp_dict)
import pandas as pd
import numpy as np

df = pd.DataFrame(temp_list)
df.head()
df.to_csv("muckrack_media_fetchlist.csv")
```

Listing 7-20 shows how to extract useful information from Muck Rack media profiles using beautifulsoup. We have been very careful to only send a handful of randomized requests by specifying the number of pages to be fetched as a function parameter. Muck Rack media profiles include lots of useful information such as geographical coverage scope, language, country of origin, social media handles, website URLs, and description, and probably most importantly, they also include a list of journalists associated with a particular publication. We have simply loaded journalists' names and profiles as a JSON so that we can denest them only when required.

Listing 7-20. Parsing media profiles from muckrack.com

```
import json
import random
import time

def parse_muckrack_media(sitemap_df, number_of_pages):
    final_list = []
    random_int_list = []

    for i in range(number_of_pages):
        random_int_list.append(random.randint(0, len(df)))

    while len(random_int_list) != 0:
        url_index = random_int_list.pop()
        url = sitemap_df.url.iloc[url_index]
        time.sleep(5)
        r = requests.get(url = url, headers = my_headers)
        html_source = r.text
        soup = BeautifulSoup(html_source, 'html.parser')
```

```python
temp_dict = {}
temp_dict["muckrack_profile_url"] = url

try:
    temp_dict["source_name"] = soup.find('h1', {'class': "mr-font-
    family-2 top-none bottom-xs"}).get_text()
except:
    temp_dict["source_name"] = ''

try:
    temp_dict["description"] = soup.find('div', {'class', 'top-
    xs'}).get_text()
except:
    temp_dict["description"] = ''

try:
    temp_dict["media_type"] = soup.find('div',{'class':'mr-font-
    weight-semibold'}).get_text()
except:
    temp_dict["media_type"] = ''

try:
    temp_dict["url"] = soup.find('div', {'class' : 'mr-contact-
    item-inner '}).get_text()
except:
    temp_dict["url"] = ''
try:
    temp_dict["twitter"] = soup.find('a',{'class', 'mr-contact
    break-word top-xs js-icon-twitter mr-contact-icon-only'})
    ['href']
except:
    temp_dict["twitter"] = ''
try:
    temp_dict["linkedin"] = soup.find('a', {'class', 'mr-contact
    break-word top-xs js-icon-linkedin mr-contact-icon-only'})
    ['href']
```

```
        except:
            temp_dict["linkedin"] = ''
        try:
            temp_dict['facebook'] = soup.find('a', {'class', 'mr-contact
            break-word top-xs js-icon-facebook mr-contact-icon-only'})
            ['href']
        except:
            temp_dict['facebook'] = ''
        try:
            temp_dict['youtube'] = soup.find('a', {'class', 'mr-contact
            break-word top-xs js-icon-youtube-play mr-contact-icon-only'})
            ['href']
        except:
            temp_dict['youtube'] = ''
        try:
            temp_dict['Pinterest'] = soup.find('a', {'class', 'mr-contact
            break-word top-xs js-icon-pinterest mr-contact-icon-only'})
            ['href']
        except:
            temp_dict['Pinterest'] = ''
        try:
            temp_dict['Instagram'] = soup.find('a', {'class', 'mr-contact
            break-word top-xs js-icon-instagram mr-contact-icon-only'})
            ['href']
        except:
            temp_dict['Instagram'] = ''

        for tr in soup.find_all('tr'):
            tds = tr.find_all('td')
            th = tr.find_all('th')
            try:
                temp_dict[th[0].get_text().strip()] = tds[0].get_text().
                strip()
```

```
        except:
            pass
    jr_list = []

    bottom_section = soup.find_all("div", {'class', 'row bottom-sm'})

    rows = soup.find_all('div', {'class', 'mr-directory-item'})
    jr_list = []
    for row in rows:
        if row is not None:
            jr_dict = {}
            jr_dict["name"] = row.get_text().strip()
            jr_dict["profile_url"] = 'https://muckrack.com'+row.
            find('a')["href"]
            jr_list.append(jr_dict)
    temp_dict["journalists"] = json.dumps(jr_list)
    final_list.append(temp_dict)
    return final_list
sample_list = parse_muckrack_media(df, 5)
df_sample = pd.DataFrame(sample_list)
df_sample.to_csv("muckrack_media.csv", index = False)
df_sample.head()
# Output
```

	Country	Days Published	Frequency	Instagram	Language	Pinterest	Scope	UVM Insights by	description	facebook	journalists
0	United Kingdom	NaN	NaN	https://instagram.com /vivamagazines	English		Local	Request pricing	VIVA Magazine is a exciting addition to the wo...	https://www.facebook.com /vivamagazines/	[{"name": "Bones, Lauren", "profile_url": "htt...	https:/. /com
1	Spain	Mon, \nTue, \nWed, \nThu, \nFri, \nSat, \nSun	Daily		Spanish		Local	Request pricing	The news must be told as always, on paper, and...	https://www.facebook.com /latribunacuenca/?ref=...	[]
2	Mexico	NaN	NaN	https://instagram.com /must_magazine	Spanish		National	Request pricing		https://www.facebook.com /MustTechStyle	[]
3	United States of America	NaN	Biyearly	https://www.instagram.com /cakeboymag/	English		International, \n \n ...	Request pricing	Cakeboy magazine is a bi-annual print publicat...	https://www.facebook.com /CakeboyMag/	[]
4	United States of America	Thu	Weekly		English		Consumer	Request pricing	We are such big fans of the Fixer Upper TV Sho...	https://www.facebook.com /groups /FixerUpperPodc...	[]

You can build a similar database for authors by scraping from muckrack.com's person profiles. Now that we have all the supporting data for a production-ready sentiment analysis model such as domain authority, media outlet details, author details, and so on, let's tackle getting millions of news articles into our database by using a distributed computing architecture in the next section.

Introduction to distributed computing

I think by this point, you have a very good idea of working with web crawl data, but we still have not tackled how to do it efficiently and quickly. We will write a couple of scripts in this section which will process news articles from the common crawl dataset using a distributed computing framework.

One of the easiest ways to process data faster is by running multiple servers on the cloud such as AWS with each processing data independent of other servers.

Figure 7-1 shows a simple distributed computing architecture with individual steps described as follows.

Step 1: Fill the Simple Queue Service (SQS) queue with tasks by the main server, which could simply be your local computer. There can be a maximum of 120,000 SQS messages in memory without being processed, known as "in-flight messages." You should spin up a number of worker instances to ensure that your in-flight messages never exceed this number.

Step 2: A Python script running on multiple EC2 servers (called worker) will request a message from SQS.

Step 3: It downloads the relevant common crawl dataset file from S3 and performs some processing steps.

Step 4: Upload the processed dataset to S3.

Step 5: The main server starts downloading the processed data from S3 and, after any data wrangling (if necessary), initiates inserting the data to the database such as Amazon RDS's PostgreSQL discussed in Chapter 5.

We have only shown the minimum number of elements necessary to process a sizable fraction of a common crawl dataset. You will still have to start and stop EC2 workers manually for simplicity, but it could be easily automated by initiating the workers through the main server and stopping the workers by sending a Simple Notification Service (SNS) message which can be used as a trigger to initiate an AWS Lambda function which stops the workers once SQS is empty.

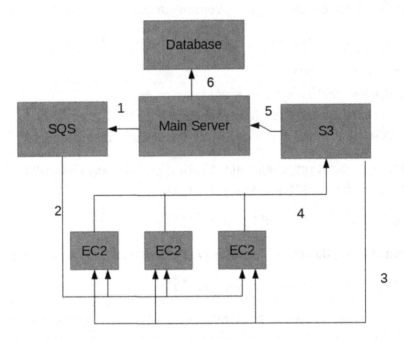

Figure 7-1. *Distributed computing architecture*

Filling an SQS queue with messages can be handled manually through your primary server which can serve as a "main" node; the other servers could be the ones located on EC2 in the US-East-1 region so that we have the lowest latency to download raw web crawls from common crawl's S3 bucket.

When you are running the script the first time, you will need to create a new SQS queue shown in Listing 7-21 to handle this task.

Listing 7-21. Creating an SQS queue

```python
import boto3
import json
import sys
import time

def CreateQueue(topic_name):

        sqs = boto3.client('sqs',  region_name = 'us-east-1')

        millis = str(int(round(time.time() * 1000)))

        #create SQS queue
        sqsQueueName=topic_name + millis
        sqs.create_queue(QueueName=sqsQueueName)
        sqsQueueUrl = sqs.get_queue_url(QueueName=sqsQueueName)['QueueUrl']

        attribs = sqs.get_queue_attributes(QueueUrl=sqsQueueUrl,
        AttributeNames=['QueueArn'])['Attributes']

        sqsQueueArn = attribs['QueueArn']

        return({"sqsQueueArn":sqsQueueArn,"sqsQueueUrl":sqsQueueUrl})

response_dict = CreateQueue("cc-news-daily")
```

Listing 7-22 fetches all the captures from theguardian.com from the March 2020 crawl. Ideally, you want to use multiple monthly web crawls using the Athena database and get news articles from thousands of top news sites too like WSJ, CNN, and so on.

Listing 7-22. Fetching JSON through the cc-index API

```
import urllib

def get_index_url(query_url):
    query = urllib.parse.quote_plus(query_url)
    base_url = 'https://index.commoncrawl.org/CC-MAIN-2020-16-index?url='
    index_url = base_url + query + '&output=json'
    return index_url
query_url = 'theguardian.com/*'
index_url = get_index_url(query_url)

import re
import time
import gzip
import json
import requests
try:
    from io import BytesIO
except:
    from StringIO import StringIO
def get_index_json(index_url):
    pages_list = []

    for i in range(4):
        resp = requests.get(index_url)
        print(resp.status_code)

        time.sleep(0.2)

        if resp.status_code == 200:
            for x in resp.content.strip().decode().split('\n'):
                page = json.loads(x)

                try:
                    if page['status'] == '200':
                        pages_list.append(page)
```

```
                except:
                    pass

            break
    return pages_list

index_json = get_index_json(index_url)
print(len(index_json))
# Output
7107
```

Let's fill this SQS queue with all the results from index_json in Listing 7-23.

Listing 7-23. Loading messages on SQS

```
response_dict = {'sqsQueueArn': 'arn:aws:sqs:us-east-1:896493407642:cc-
news-daily1597659958131',
 'sqsQueueUrl': 'https://queue.amazonaws.com/896493407642/cc-news-
daily1597659958131'}

import boto3
import json
from datetime import datetime
def myconverter(o):
    if isinstance(o, datetime):
        return o.__str__()
# Create SQS client
sqs = boto3.client('sqs',region_name = 'us-east-1')

queue_url = response_dict["sqsQueueUrl"]
for line in index_json:
    payload = json.dumps(line, default = myconverter)
    # Send message to SQS queue
    response = sqs.send_message(
        QueueUrl=queue_url,
        DelaySeconds=10,
        MessageAttributes={

        },
```

```
        MessageBody=(
            payload
        )
    )

    print(response['MessageId'])
#Output
024829f7-7847-41d3-b67d-d7e3c4dbcbcc
ab0985d9-8c8c-43e9-81e2-6321eade72d5
```

Once the SQS queue is filled with tasks, we can write a worker side script to download and parse news articles from the S3 bucket as shown in Listing 7-24.

Listing 7-24. Worker node script download and parse from the S3 bucket

```
import json
from newspaper import Article
import numpy as np
import pandas as pd

def newspaper_parse(html):

    article = Article('')
    article.set_html(html)
    article.download()

    article_title = None
    json_authors = None
    article_text = None
    article_publish_date = None

    try:

        article.parse()

        json_authors = json.dumps(article.authors)

        article_title = article.title
        article_text = article.text
        article_publish_date = article.publish_date
```

```python
    except:

        pass

    return  article_title, json_authors, article_text, article_publish_date
def get_html_from_cc_index(page):

    offset, length = int(page['offset']), int(page['length'])
    offset_end = offset + length - 1
    prefix = 'https://commoncrawl.s3.amazonaws.com/'
    temp_list = []

    try:

        resp2 = requests.get(prefix + page['filename'], headers={'Range':
        'bytes={}-{}'.format(offset, offset_end)})

        raw_data = BytesIO(resp2.content)
        f = gzip.GzipFile(fileobj=raw_data)
        data = f.read()
    except:

        print('some error in connection?')

    try:

        temp_dict = {}
        warc, header, response = data.strip().decode().split('\r\n\r\n', 2)
        temp_dict["article_title"], authors_list, temp_dict["article_
        text"], temp_dict["article_publish_date"] = newspaper_
        parse(response)
        temp_dict["url"] = page["url"]
        authors_list = json.loads(authors_list)
        if len(authors_list) == 0:
            temp_dict["author"] = ''
            temp_list.append(temp_dict)
```

```
        else:
            for author in authors_list:

                temp_dict["author"] = author

    except Exception as e:
        pass
        print(e)

    return temp_dict
```

Listing 7-25 shows how to fetch SQS messages and create files of about 1000 rows and upload them back to S3. We are limiting the row size of each file for using the server memory optimally and also making it easier to process the files later by the main server.

Listing 7-25. Iterating through the SQS queue

```
import pandas as pd
import numpy as np
import os
import uuid

def upload_to_s3(final_list, S3_bucket_name):

    local_filename = str(uuid.uuid4()) + '.csv'

    df = pd.DataFrame(final_list)
    df.to_csv(local_filename, index = False)

    s3 = boto3.client('s3',region_name = 'us-east-1')

    for attempt in range(1,6):
        try:
            # files automatically and upload parts in parallel.
            s3.upload_file(local_filename,S3_bucket_name, local_filename)

        except Exception as e:
            print(str(e))
        else:
            print("finished uploading to s3 in attempt ", attempt)
            break
```

```
        os.remove(local_filename)

final_list = []
while True:

    sqs = boto3.client('sqs',  region_name = 'us-east-1')

    try:
        sqsResponse = sqs.receive_message(QueueUrl=response_
        dict['sqsQueueUrl'], MessageAttributeNames=['ALL'],
        MaxNumberOfMessages=1, WaitTimeSeconds = 10)

        page = json.loads(sqsResponse["Messages"][0]["Body"])

        receipt_handle = sqsResponse["Messages"][0]["ReceiptHandle"]
        response = sqs.delete_message(QueueUrl=response_
        dict['sqsQueueUrl'], ReceiptHandle=receipt_handle)

        final_list.append(get_html_from_cc_index(page))

        if len(final_list) == 1000:
            upload_to_s3(final_list, 'ec2-testing-for-s3-permissions')
            final_list = []

    except Exception as E:
        print('no more messages to fetch')
        upload_to_s3(final_list, 'ec2-testing-for-s3-permissions')
        break
# Output
no more messages to fetch
finished uploading to s3 in attempt  1
```

Listing 7-26 shows the dataframe with parsed content from theguardian.com. We can easily include scripts discussed in Chapter 4 to also scrape email addresses from the web page and perform topic classification.

Listing 7-26. Parsed content from theguardian.com

```
df_responses = pd.DataFrame(final_list)
df_responses.head()
#Output
```

	article_publish_date	article_text	article_title	author	url
0	NaT	Katharine Viner is editor-in-chief of the Guar...	About the Guardian		https://www.theguardian.com/about
1	2010-06-15 09:39:34+00:00	What term do you want to search? Search with g...	Video interview with Rose Shuman, founder, Ope...		https://www.theguardian.com/activate/video-int...
2	2011-05-27 09:24:00+00:00	What term do you want to search? Search with g...	Activate New York: Rose Shuman - video		https://www.theguardian.com/activate/video/act...
3	NaT	US Climate Alliance, Climate Mayors, We Are St...	Advertiser content hosted by the Guardian: The...		https://www.theguardian.com/advertiser-content...
4	2010-06-15 09:39:34+00:00	What term do you want to search? Search with g...	Video interview with Rose Shuman, founder, Ope...		https://www.theguardian.com/activate/video-int...

You may have noticed that we are only running Python scripts in a single process; so we are not utilizing computing power from all the available CPU cores of the server. In Listing 7-27, we have two scripts; the first is the multiprocessing_individual.py which simply combines the code in Listings 7-24 and 7-25 into a single script. The other script is the multiprocessing_main.py which initiates multiprocessing pool according to the CPU count of the server.

Listing 7-27. Multiprocessing example

```python
# save it in a file named multiprocessing_main.py

from multiprocessing import Pool
import multiprocessing
import os
def run_process(process):
        os.system('python {}'.format(process))
if __name__ == '__main__':

    sample_file_path = 'multiprocessing_individual.py'

    #print(sample_file_path)
    processes = []
    pool_count = multiprocessing.cpu_count()
    print("cpu pool count is " + " " + str(pool_count))
    for item in range(pool_count):
        processes.append(str(sample_file_path))
    processes = tuple(processes)

    #logging.info("pooled processes started")
    pool = Pool(pool_count)
    pool.map(run_process, processes)
    # Note to Reader: add code here to shut off EC2
```

We should also set up the multiprocessing_main.py script via crontabs so that every time the EC2 server starts, it will trigger the multiprocessing main script, which in turn triggers the multiple multiprocessing_individual.py scripts according to the number of CPU cores. Let's recall that the script in Listing 3-12 was configured to start automatically whenever we start an instance of EC2 by using crontabs.

Now, if the SQS queue is full, all the individual processes will start fetching messages from SQS and downloading the relevant files from S3 and pushing out a parsed output file with 1000 rows or at the end of the queue. Now, once the SQS queue gets empty, the individual process stops and exits the while loop. Once all the processes are stopped, the multiprocessing_main.py script itself exits. I would recommend that you should check out AWS Lambda and use it to shut down EC2 instances by simply using a Simple Notification Service (SNS) message as a trigger. If you set that up in the last line of multiprocessing_main.py, then the worker servers will shut off once the SQS queue is empty without any manual input.

There are countless other more advanced ways to perform distributed computing, but I just wanted to discuss a method with a low learning curve.

Rolling your own search engine

So at this point, you might be wondering how easy is it to roll your own search engine using web crawls? A PostgreSQL and Athena–based web crawl search will never give you query times comparable to traditional web search engines, and hence you will have to look at Elasticsearch, CloudSearch, Solr, and so on.

There is a publicly accessible, open source project called Elastic ChatNoir (`https://github.com/chatnoir-eu`).

If you are serious at crawling a large fraction of the Web regularly, then I highly recommend using a Java-based stack to do it with a great starting point being the common crawl codebase (`https://github.com/commoncrawl`). Python is good for many things, but it still doesn't have a comparable full text search library such as Apache Lucene on the top of which Elasticsearch is built. Similarly, there is no production-ready broad crawler comparable to Apache Nutch (used by the common crawl project) in the Python ecosystem. Lucene, Nutch, and Hadoop are tightly linked together since their creator Doug Cutting designed them primarily for crawling and indexing the Web over 15 years ago (`https://queue.acm.org/detail.cfm?id=988408`), and they have stood the test of time.

The challenge of building a free-to-use publicly available search engine is not technological at all, but a purely financial one. It will cost thousands of dollars a month in server charge to replicate Elastic ChatNoir, and trying to monetize this expense via ads is extremely difficult due to a complete lock of the search engine market by Google and Bing.

There have been few people who have made a go at this in the last few years, with a notable mention of Blekko which ran a search engine for many years before selling it to IBM for the Watson project in 2015.

DuckDuckGo has been one of the most promising stories in the publicly available, free-to-use, commercial search engine space that has been able to compete against established players like Google by monetizing on ads while still being a relatively small company of just about 100 employees and still headquartered in Paoli, PA (United States).

However, this is by far an exception to the rule, and for most data-centric companies such as Ahrefs, Moz, and Hunter and including us at Specrom Analytics, it makes more sense to not even contemplate providing a publicly available, unrestricted use search engine but rather make results available from web crawls via public APIs such as the latest news API (`https://algorithmia.com/algorithms/specrom/LatestNewsAPI`), email address search API (`https://algorithmia.com/algorithms/specrom/Get_email_addresses_by_domain`), and many others with metering with generous free-tier usage so at least we can regulate the load on our servers.

Summary

We learned about how to use Amazon Athena to directly query data located in the S3 bucket and used it for processing the common crawl index and querying for domain authority and ranking.

We also revisited sentiment analysis and went through all the different types of data to make it comparable to data from commercial providers.

Lastly, we discussed a simple distributed computing framework to process web crawl data on a big data scale.

This chapter wraps up all the important use cases for web scraping that we had talked about in Chapter 1.

In the next chapter, we'll focus our attention to running focused crawlers on scale for scraping information from web domains with aggressive antiscraping measures such as Amazon.com by using IP rotation, user-agent rotation, CAPTCHA solving service, and so on.

CHAPTER 8

Advanced Web Crawlers

In this chapter, we will discuss a crawling framework called Scrapy and go through the steps necessary to crawl and upload the web crawl data to an S3 bucket.

We will also talk about some of the practical workarounds for common antibot measures such as proxy IP and user-agent rotation, CAPTCHA solving services, and so on.

Scrapy

Scrapy is a very popular production-ready web crawling framework in Python; it contains all the features of a good web crawler such as robots.txt parser, crawl delay, and Selenium support that we talked about in Chapter 2 right out of the box.

Scrapy might prove a bit tricky to install with just PIP on your computer since you will have to take care of all the third-party dependencies yourself; a much better idea is to install it using conda:

```
conda install -c conda-forge scrapy
```

Scrapy abstracts away lots of low-level details for operating a crawler. We will create a crawler very similar to Listing 2-14 of Chapter 2 that crawled through the pages of my personal website.

Once it is installed, initiate a new Scrapy project by typing

```
scrapy startproject chapter_8
```

Now, initiate the first spider by entering the chapter_8 directory (cd chapter_8) and running the following commands:

```
scrapy genspider linkscraper-basic jaympatel.com
scrapy genspider second-scraper jaympatel.com
```

© Jay M. Patel 2020
J. M. Patel, *Getting Structured Data from the Internet*, https://doi.org/10.1007/978-1-4842-6576-5_8

This should give you a directory structure shown in Figure 8-1; let's call the base directory scrapy_home.

```
scrapy.cfg
chapter_8
    items.py
    middlewares.py
    pipelines.py
    settings.py
    __init__.py

    spiders
        linkscraper_basic.py
        second_scraper.py
        __init__.py
```

Figure 8-1. *Scrapy directory structure*

Scrapy is really not suited for the Jupyter Notebook code format we have been using in supporting information so far; hence, you will see the code meant for Scrapy in the Jupyter Notebook for this chapter, but you will have to copy-paste it to the appropriate .py file in the Scrapy directory.

Let's look at the settings.py file; it will consist of only three lines of code as shown in Listing 8-1 with the rest being commented out. You can delete all the commented out sections to avoid any confusion.

Listing 8-1. Default settings.py file contents

```
BOT_NAME = 'chapter_8'
SPIDER_MODULES = ['chapter_8.spiders']
NEWSPIDER_MODULE = 'chapter_8.spiders'
ROBOTSTXT_OBEY = True
```

We notice right away that Scrapy has abstracted away the low-level coding for parsing robots.txt like we had to do in Listing 2-16 of Chapter 2. Let's add some more parameters to the settings.py file shown in Listing 8-2. The user agent string mentioned here is the same as the one we used in requests objects in Listing 2-7; Scrapy is just letting us abstract that away from our application code.

concurrent_requests, as the name indicates, specifies the number of requests Scrapy can make concurrently. In production, we use a number closer to 100, but since we are testing this on my personal website, I specified one concurrent request.

download_delay simply specifies how long Scrapy should wait (in seconds) before requesting web pages from the same domain address. Recall back to our discussion in Chapter 2 about crawl delay and about the importance of setting a reasonable delay between requests to avoid getting blocked completely. Scrapy also provides more advanced methods of setting a crawl delay using an autothrottle extension which sets the crawl delay dynamically based on loading speeds of the scraped website.

download_timeout sets the maximum time to wait when requesting a web page before timing out a request. This is very necessary when trying to crawl a large number of domains. If we were implementing this on our own, then we will specify timeout using a requests module like this:

```
r = requests.get(url, timeout=15)
```

redirect_enabled takes in a boolean value for enabling redirects. We are leaving it True here, but you should change it depending on your end use.

In Chapter 2, we learned about crawl depth and crawl order which simply determine the exact order in which a crawler discovers new links. As a default, Scrapy performs a depth-first crawling using a last-in-first-out (LIFO) queue. Instead of that, it's a much better idea to perform a breadth-first search (BFS) where the crawler first discovers all the links on a given page and then proceeds to crawl it in a first-in-first-out (FIFO) queue systematically in case the depth level meets a predetermined threshold. We can specify a depth level and a BFS search by specifying the depth_limit, depth_priority, scheduler_disk_queue, and scheduler_memory_queue.

Listing 8-2. Additional settings.py contents

```
USER_AGENT = 'Mozilla/5.0 (Windows NT 10.0; Win64; x64) AppleWebKit/537.36
(KHTML, like Gecko) Chrome/61.0.3163.100Safari/537.36'

CONCURRENT_REQUESTS = 1

DOWNLOAD_DELAY = 0.05

DOWNLOAD_TIMEOUT = 15

REDIRECT_ENABLED = True
```

```
DEPTH_LIMIT = 3
DEPTH_PRIORITY = 1
SCHEDULER_DISK_QUEUE = 'scrapy.squeues.PickleFifoDiskQueue'
SCHEDULER_MEMORY_QUEUE = 'scrapy.squeues.FifoMemoryQueue'

LOG_LEVEL = 'INFO'
```

Let's turn our attention to the items.py file in the spiders folder shown in Listing 8-3. items.py allows us to specify data fields we want to extract out from a web page.

Listing 8-3. items.py default contents

```
# -*- coding: utf-8 -*-

# Define here the models for your scraped items
#
# See documentation in:
# https://doc.scrapy.org/en/latest/topics/items.html

import scrapy


class Chapter8Item(scrapy.Item):
    # define the fields for your item here like:
    # name = scrapy.Field()
    Pass
```

I will keep things simple for now and only specify three fields for the URL, depth, and title as shown in Listing 8-4.

Listing 8-4. items.py fields

```
import scrapy

class Chapter8Item(scrapy.Item):
    url = scrapy.Field()
    title = scrapy.Field()
    depth = scrapy.Field()
```

Lastly, we will look at the linkscraper_basic.py scraper located in the spiders folder as shown in Listing 8-5. We notice right away that it's already prepopulated with domain information collected when we created the spider using the scrapy genspider command. As a default, the crawl is being restricted to only one domain from the allowed_domains list, and the seed URL is the domain homepage. Scrapy supports advanced rules based on regular expressions to specify which pages you would like to scrape in minute granularity using crawlspider class rules (`https://docs.scrapy.org/en/latest/topics/spiders.html#crawlspider`). We will only crawl through jaympatel.com so that we can compare the results directly with Listing 2-14.

Listing 8-5. linkscraper_basic.py default contents

```
# -*- coding: utf-8 -*-
import scrapy

class LinkscraperBasicSpider(scrapy.Spider):
    name = 'linkscraper-basic'
    allowed_domains = ['jaympatel.com']
    start_urls = ['http://jaympatel.com/']

    def parse(self, response):
        pass
```

Scrapy provides its own data selectors to parse data out of the web page's response object. Instead of using them, I advise you to continue using Beautifulsoup and lxml libraries we saw in Chapter 2 for HTML parsing so that you can make your code more portable and maintainable across web crawlers. After all, Scrapy selectors are wrappers around the parsel library (`https://docs.scrapy.org/en/latest/topics/selectors.html`) that uses the lxml library under the hood, so it makes sense to forgo some convenience in favor of speed and maintainability by calling the underlying library directly.

Scrapy will start parsing by calling the start_url and sending the response to the parse function. Our parse function will extract titles from the response and save those as well as the URL and depth in the items object we created in Listing 8-4. We will also iterate through all the URLs discovered on the page and use the response.follow() method which will automatically put the URL in crawl queue if it's from this same domain, and it also takes care of filling and adding the base domain to the relative URL path.

Listing 8-6. Complete linkscraper_basic.py function

```python
# -*- coding: utf-8 -*-
import scrapy
from bs4 import BeautifulSoup
from chapter_8.items import Chapter8Item

class LinkscraperBasicSpider(scrapy.Spider):
    name = 'linkscraper-basic'
    allowed_domains = ['jaympatel.com']
    start_urls = ['http://jaympatel.com/']

    def parse(self, response):

            item = Chapter8Item()
            if response.headers["Content-Type"] == b'text/html;
            charset=utf-8' or response.headers["Content-Type"] == b'text/
            html':
                soup = BeautifulSoup(response.text,'html.parser')
                urls = soup.find_all('a', href=True)
                for val in soup.find_all('title'):
                    try:
                        item["url"] = response.url
                        item["title"] = val.get_text()
                        item["depth"] = str(response.meta['depth'])
                        yield item
                    except Exception as E:
                        print(str(E))

            else:
                item["title"] = 'title not extracted since content-type is
                ' + str(response.headers["Content-Type"])
                item["url"] = response.url
                item["depth"] = str(response.meta['depth'])
                urls = []
                yield item

            for url in urls:
                yield response.follow(url['href'], callback=self.parse)
```

Now, all we have to do is call the command shown as follows from scrapy_home to crawl through the website and export the contents to a JSON Lines file.

```
scrapy crawl linkscraper-basic -o pages.jl
```

You should see a new file called pages.jl in the scrapy_home directory. We will explore the file contents in our Jupyter Notebook as shown in Listing 8-7.

Listing 8-7. Exploring the pages.jl file

```
import json
file_path = 'pages.jl'

contents = open(file_path, "r").read()
data = [json.loads(str(item)) for item in contents.strip().split('\n')]
for dd in data:
    print(dd)
    print("*"*10)
#Output
{'url': 'http://jaympatel.com/', 'title': 'Jay M. Patel', 'depth': '0'}
**********
{'url': 'http://jaympatel.com/tags/', 'title': 'Jay M. Patel', 'depth':
'1'}
**********
{'url': 'http://jaympatel.com/2019/02/using-twitter-rest-apis-in-python-
to-search-and-download-tweets-in-bulk/', 'title': '\n    Using Twitter rest
APIs in Python to search and download tweets in bulk - Jay M. Patel\n',
'depth': '1'}
... (output truncated)..
{'title': "title not extracted since content-type is b'application/pdf'",
'url': 'http://jaympatel.com/pages/CV.pdf', 'depth': '3'}
**********
{'title': "title not extracted since content-type is b'application/pdf'",
'url': 'http://jaympatel.com/assets/DoD_SERDP_case_study.pdf', 'depth': '3'}
**********
```

So we notice that the crawl worked pretty well, and it also stopped at the correct depth as specified in the settings.py file. We could have also used other file formats such as CSV and JSON which are available as a default with Scrapy; however, JSON Lines file format allows you to write to a file as a stream, and it's pretty popular for handling crawls with thousands of pages.

Scrapy pipelines and Scrapy middlewares also let you build custom pipelines for exporting the data directly into your database. Let us install a third-party package called s3pipeline (pip3 install scrapy-s3pipeline) which lets us upload the JSON Lines file directly to the S3 bucket of our choice.

You will have to edit the settings.py file to add the following parameters shown in Listing 8-8. The item_pipelines parameter simply activates the S3 pipeline; the S3pipeline_url is formatted to contain S3://bucket_name/folder_name/. The individual filename for the JSON Lines file will be of the format {time}.{chunk:07d}.jl.gz. The max chunk size specifies the length of each file; we have seen that each warc file in Chapter 6 contained raw responses from around 20,000–50,000 web pages and so should probably stay at or below that level or the files will get too large to process effectively. s3pipeline_gzip is a boolean parameter which specifies if we want a compressed file or not.

If your computer already has AWS credentials configured via AWS CLI, then you do not have to specify the remaining parameters. However, if your default region is different from the location of the S3 bucket, then you can just specify that here.

Listing 8-8. Additional settings.py parameters for using s3pipeline

```
ITEM_PIPELINES = { 's3pipeline.S3Pipeline': 100}

S3PIPELINE_URL = 's3://athena-us-east-1-testing/chapter-8/{time}.
{chunk:07d}.jl.gz'

S3PIPELINE_MAX_CHUNK_SIZE = 10000

S3PIPELINE_GZIP = True

# If different than AWS CLI configure values

AWS_REGION_NAME = 'us-east-1'

AWS_ACCESS_KEY_ID = 'YOUR_VALUE'
AWS_SECRET_ACCESS_KEY = 'YOUR_VALUE'
```

Now, all we need to do is test out our new pipeline by entering the following command from the scrapy_home directory:

```
scrapy crawl linkscraper-basic
```

We will get a file in the S3 folder; let's explore it in Listing 8-9 by downloading it to ensure that it matches the one we saw in Listing 8-7 (it does!). You could have downloaded the file using Boto3 directly from S3 and opened it in the Jupyter Notebook; I left out that code as an exercise since it's something we have already done many times before in Chapters 3 and 6.

Listing 8-9. jl.gz output from the S3 folder

```
import gzip
import json
file_path_gzip= 'FILENAME_ON_S3.jl.gz'

data = []
with gzip.open(file_path_gzip,'r') as fin:

    for item in fin:
        #print('got line', data.append(json.loads(item)))
        data.append(json.loads(item))
for dd in data:
    print(dd)
    print("*"*10)
```

Let us take a step back from Scrapy to look at the overall web crawling strategy itself. Most Scrapy users are trying to perform what we commonly refer to as "focused crawling," which is simply fetching content from a narrow list of web pages which fulfill a set of specific conditions. In its simplest sense, it may just mean that we are only fetching web pages from a specific domain just like we did earlier.

We could also create focused crawlers by using a broader set of rules such as crawling only pages from domains with a domain ranking of, say, 10,000 or lower. In Chapter 6, we introduced similarity scores, and these can also be used for focused crawling by taking into account the similarity of anchor text with a gold standard and only fetching the page if it's higher than a certain threshold. All of these cases should be easy enough for you to implement based on what you have learned already in the previous chapters.

When you are performing focused crawling on a relatively small domain with only a few thousand pages, it will make near zero difference performance wise if we were to combine crawling with parsing of the web page like we did in Listing 8-6.

However, as the size of the crawls increases, and once they start taking hundreds of hours to finish a given crawl job, then you will make significant savings if you simply save raw web pages during crawling and parse them later in a separate distributed workflow. For crawling to happen, we still need to perform a basic links discovery level of parsing, but that is pretty computationally light compared to traversing the HTML tree entirely or running expensive natural language processing (NLP) algorithms.

We run web parsers at Specrom Analytics on spot EC2 instances which can be 30–50% cheaper than using on-demand instances.

One major issue with spot instances is that they can be shut down anytime by AWS so you cannot really run Scrapy with long-running crawl tasks on them; but you can use them for parsing raw web pages based on some SQS queue, so even if they get shut down, other instances can take over and complete the job.

The other reason for separating out parsing and crawling aspects is the ability to use the same raw web crawls for multiple applications listed in Chapter 1. It will be cost prohibitive to extract full text and run natural language processing (NLP) algorithms such as NER (named entity recognition) and technology profiling and create a backlinks database on all your web crawls. It is rather much better if you have raw web crawls stored somewhere like S3 Glacier and only analyze the aspects your products need now or for any services you are providing to your clients. In essence, what I am saying is that there are too many possible applications for raw web crawls that you should save them periodically and only analyze parts of them as you need.

Lastly, a website schema changes over time; this means that you need to update the parsing logic way more often than the base crawler itself, so it's better to separate out the two and perform the updates in different codebases.

So let us alter our items.py file to only capture three things from our crawls: the URL, response headers, and raw response itself as shown in Listing 8-10. These are very similar to the ones we saw in common crawl's WARC files in Chapter 6.

Listing 8-10. Modifying the items.py file to capture raw web crawl data

```python
class Chapter8ItemRaw(scrapy.Item)
    headers = scrapy.Field()
    url = scrapy.Field()
    response = scrapy.Field()
    crawl_date = scrapy.Field()
```

Lastly, let's also insert code into the second-scraper we had created initially in the spiders folder in Listing 8-11. We have used the linkextractors to extract links from the page and continue the crawl.

Listing 8-11. second-scraper.py

```python
# -*- coding: utf-8 -*-
import scrapy
from datetime import datetime, timezone
from scrapy.linkextractors import LinkExtractor
from chapter_8.items import Chapter8ItemRaw

class SecondScraperSpider(scrapy.Spider):
    name = 'second-scraper'
    allowed_domains = ['jaympatel.com']
    start_urls = ['http://jaympatel.com/']

    def parse(self, response):

            item = Chapter8ItemRaw()
            item['headers'] = str(response.headers)
            item['url'] = response.url
            item['body'] = response.text
            item['crawl_date'] = datetime.now(timezone.utc).
            replace(microsecond=0).isoformat()
            yield item

            for a in LinkExtractor().extract_links(response):
                yield response.follow(a, callback=self.parse)
```

Let's call this scraper by entering

```
scrapy crawl second-scraper
```

Listing 8-12 shows the file containing the raw web crawls, and unsurprisingly it's the same result as Listing 8-7 except that the parsing is being done here itself.

Listing 8-12. Parsing jl.gz containing raw web crawls

```
import gzip
import json
from bs4 import BeautifulSoup

file_path_gzip= 'FILENAME_ON_S3.jl.gz'

data = []
with gzip.open(file_path_gzip,'r') as fin:
        for item in fin:
                data.append(json.loads(item))
for dd in data:
    print(dd["url"])
    #print(dd["headers"])
    soup = BeautifulSoup(dd["response"],'html.parser')
    print(soup.find('title').get_text())
    print("*"*10)
```

So how can we scale up Scrapy to crawl on a large scale? Unfortunately, it's not really designed for broad crawling, but you can modify some settings listed here (`https://docs.scrapy.org/en/latest/topics/broad-crawls.html`) to make it more suitable for it. If you are directly comparing it to a broad crawler such as Apache Nutch, then Scrapy might not meet your expectations since it doesn't integrate directly with Hadoop and support distributed crawling right out of the box.

An easy way to scale up is by simply vertical scaling where you run Scrapy on a powerful server so that it can fetch more pages concurrently; this coupled with IP rotation strategies discussed in the next section should be sufficient in many cases. Another potential method is running Scrapy independently on separate servers and restricting each instance to only crawl a predefined domains list. Lastly, you can also check out Frontera (`https://frontera.readthedocs.io/en/latest/`) which is tightly integrated with Scrapy and might better suit your needs. I will reiterate the point I made at the end of Chapter 7 about checking out Java-based crawlers such as Apache Nutch in case your workload is truly exceeding Scrapy's capabilities. A good starting point is using common crawl's fork of Apache Nutch 1.x–based crawler. The codebase is open sourced and available on the GitHub repo (`https://github.com/commoncrawl`), and it's not only well documented but also stable enough for production use.

Advanced crawling strategies

In this section, let us consider crawling on website domains where web crawlers such as common crawl were unable to fetch a web page by getting blocked via a CAPTCHA or some other method. Let's illustrate this by fetching the number of captures for Amazon. com on common crawl's March 2020 index as shown in Listing 8-13.

Listing 8-13. Fetching Amazon.com captures through the cc-index API

```
import urllib

def get_index_url(query_url):

    query = urllib.parse.quote_plus(query_url)
    base_url = 'https://index.commoncrawl.org/CC-MAIN-2020-16-index?url='
    index_url = base_url + query + '&output=json'
    return index_url
query_url = 'amazon.com/*'
index_url = get_index_url(query_url)

import re
import time
import gzip
import json
import requests
try:
    from io import BytesIO
except:
    from StringIO import StringIO
def get_index_json(index_url):
    pages_list = []
    #payload_content = None
```

```
    for i in range(4):
        resp = requests.get(index_url)
        #print(resp.status_code)

        time.sleep(0.2)

        if resp.status_code == 200:
            for x in resp.content.strip().decode().split('\n'):
                page = json.loads(x)

                try:

                    pages_list.append(page)

                except:
                    pass

            break
    return pages_list

index_json = get_index_json(index_url)
print(len(index_json))
# output
13622
```

We see that common crawl fetched about 13,000 pages from Amazon.com. Let's explore the status codes related to these pages in Listing 8-14.

Listing 8-14. Exploring status codes for Amazon.com page captures

```
import numpy as np
import pandas as pd

df = pd.DataFrame(index_json)
df.status.value_counts()
# Output
503     6753
301     5274
200      897
302      635
404       58
400        5
Name: status, dtype: int64
```

We see that less than 10% of the pages were fetched with a status code of 200. The rest of them had redirects (301/302) or server side errors (503). There were also some client-side errors (404/400), but they are too insignificant to matter much. Let's check out a page with a server-side error in Listing 8-15.

Listing 8-15. Page with a 503 status code

```python
page = df[df.status == '503'].iloc[1].to_dict()
import re
import time
import gzip
import json
import requests
try:
    from io import BytesIO
except:
    from StringIO import StringIO

def get_from_index(page):

    offset, length = int(page['offset']), int(page['length'])
    offset_end = offset + length - 1
    prefix = 'https://commoncrawl.s3.amazonaws.com/'

    try:

        r = requests.get(prefix + page['filename'], headers={'Range':
        'bytes={}-{}'.format(offset, offset_end)})
        raw_data = BytesIO(r.content)
        f = gzip.GzipFile(fileobj=raw_data)
        data = f.read()

    except:

        print('some error in connection?')

    try:
        crawl_metadata, header, response = data.strip().decode('utf-8').
        split('\r\n\r\n', 2)
```

```
    except Exception as e:
        pass
        print(e)

    return crawl_metadata, header, response
crawl_metadata, header, response = get_from_index(page)
soup = BeautifulSoup(response,'html.parser')
for script in soup(["script", "style"]):
        script.extract()
print(soup.get_text())
#Output:
Robot Check

Enter the characters you see below
Sorry, we just need to make sure you're not a robot. For best results,
please make sure your browser is accepting cookies.

Type the characters you see in this image:
. . .

(Output truncated)
```

It's no surprise that common crawl got served a CAPTCHA; let's recall back to Listing 2-7 of Chapter 2 where we were served with a CAPTCHA screen when trying to scrape from Amazon.com.

As we had mentioned earlier, common crawl's crawler is compliant to the robots.txt file, and it will only fetch a page which is allowed through robots.txt. So, theoretically, it should be able to fetch all the web pages it requests (with retries) if they still exist on the server, and we should see a very high number of status 200 codes.

However, there are a handful of websites like Amazon.com that implement aggressive antiscraping measures preventing pages from being fetched even when they should technically be available as per their robots.txt.

We will go through bypassing common antibot measures on both the server side and the client side, but before we get there, let's review the ethics and legality of fetching web pages by using these methods.

Ethics and legality of web scraping

I am an engineer and not a lawyer, so please do not consider this a legal advice but rather an opinion from a practitioner. Generally speaking, web scraping is legal and ethical if it's performed by remaining compliant to the robots.txt file and terms of use of a website and by declaring yourself to be a crawler by setting a recognizable user agent string such as Googlebot, CCBot, Bingbot, and so on used by Google, Common Crawl, Bing, and so on, respectively.

It's obviously unethical to disregard robots.txt or spoof user agents, IP addresses, and so on to make it appear like you are not a crawler but a real human browser, but legally speaking, it's a bit of a gray area.

From a technical perspective, even aggressive antiscraping measures to discourage scraping cannot stop you from accessing the content completely as long as the website itself is still accessible to the public. You can use strategies described in the next section to scrape data from domains which explicitly prohibit scraping through the robots.txt file or terms of use such as LinkedIn. However, I will like to quote Peter Parker's (aka Spiderman) principle created by Stan Lee which says "with great powers comes great responsibility." In a web scraping context, I mean that being technically able to scrape from all websites shouldn't mean that you actually go out and scrape on live sites with total disregard to robots.txt and a website's terms of service (ToS).

There can be severe legal repercussions of doing so, ranging from a cease and desist letter, lawsuit, hefty fines, and more. The most notable case in the recent past is a scraping company called 3Taps being forced to settle a lawsuit with Craigslist by paying $1 million in 2015 (`https://arstechnica.com/tech-policy/2015/06/3taps-to-pay-craigslist-1-million-to-end-lengthy-lawsuit-will-shut-down/`).

The broader issue of determining the legality of scraping in violation of ToS and robots.txt is being actively litigated in court for a lawsuit filed by LinkedIn, and even though a recent court order in the United States has come down in favor of a web scraping company HiQ (`www.theverge.com/2019/9/10/20859399/linkedin-hiq-data-scraping-cfaa-lawsuit-ninth-circuit-ruling`), this may change in the future. Similarly, the EU's General Data Protection Regulation (GDPR) comes into effect if you are scraping personal data of an EU resident.

Hence, the information here is provided for educational purposes only with no implied endorsement by the author to a technique or service mentioned, and you are encouraged to check out the legality of using it in your local jurisdiction before implementing it on a live website.

Proxy IP and user-agent rotation

One of the most obvious ways a website can detect its users is by keeping track of their IP address which is used while making requests to the page. If you detect hundreds of requests from a single IP address in a short span of time, then you'll be able to reliably say that it's probably an automated scraper making the request rather than a real human user. In that case, it's probably a good idea to start blocking or throttling down the requests by serving a 501/502 server error so that you can restrict your server's resources for legitimate human users.

If you are in fact running a scraper, then it will be in your best interest to send the requests using a pool of proxy IP addresses with time delays in such a way that you are only hitting a target domain a few times a minute from one IP address. By using a strategy like this, you make your scraping activities very difficult to be recognized, and it will prevent you from getting blocked. Now, if you keep using the same set of IP addresses to make the requests, then eventually you will start getting blocked, and you'll have to rotate IP addresses to new ones.

As I had mentioned in Chapter 3, if you are using cloud computing servers like EC2, then you will automatically get a new IP address every time a new instance of the EC2 server is started, except if you explicitly request a static IP address. Similarly, using serverless applications like AWS Lambda to make requests will also ensure that the requests are made with new IP addresses, and you will be relatively insulated from blocking just on the basis of your IP address.

Hence, when using cloud computing servers, there is a good chance of at least a minority of requests getting a 200 status code response back especially during the initial block of requests even if the website domain runs an aggressive antibot strategy like Amazon. We already saw in Listing 8-14 that common crawl March 2020 contains about 890 page captures from Amazon.com with a 200 status code so at least it successfully scraped some pages.

There are plenty of IP proxy services out there such as `https://smartproxy.com/`, `www.scrapinghub.com/crawlera/`, and `https://oxylabs.io/` which will provide you hundreds of IP addresses at a fixed monthly rate varying from $99 to $300/month. The cheapest proxy IP addresses are based on a data center IP address and considered more prone to blocking. The other more expensive options are residential proxies which route requests through computers on residential networks and mobile IP addresses which route requests through mobile networks, both of which will let you scrape with an even lower blocking or CAPTCHA screens.

So once you have a pool of IP addresses from the provider, all you need to do is use it with your requests like shown in Listing 8-16 with a proxy IP address as a dict with port numbers. Since you will be using a list of IP addresses, it will be a good idea to first randomly select one from the list, create a dict, and then make the request. Most proxy IP address providers provide their own wrappers that handle not only rotating the IP addresses but also retries with a different address in case the first one fails. Scrapy has a very useful middleware called scrapy-rotating-proxies (`https://pypi.org/project/scrapy-rotating-proxies/`) that handles the low-level details of the proxy IP address rotation.

Listing 8-16. Using proxy IP addresses with requests

```
import requests

proxy_ip = {
 'http': 'http://11.11.11.11:8010',
 'https': 'http://11.11.11.11:8010',
}

my_headers = {
'User-Agent': 'Mozilla/5.0 (Windows NT 10.0; Win64; x64) AppleWebKit/537.36
' + ' (KHTML, like Gecko) Chrome/61.0.3163.100Safari/537.36'
}

r = requests.get(url, proxies=proxy_ip, headers = my_headers)
```

Let us switch our attention to user agents, as you may recall from Figure 2-4 of Chapter 2, these are the set of strings you send in your request header identifying your computer and browser. We have shown how we can use user strings to make it appear to the website domain that we are sending a request from a real computer browser instead of doing it programmatically.

This is of course very rudimentary and easy to forge; however, using a user agent from a real browser definitely improves your scraping success rate vs. using no user agent string at all.

When you send a static user agent with an IP rotation like we did in Listing 8-16, you are raising a warning flag on being a bot. Think about it; what are the odds that a real human with the same exact user agent is requesting pages 1 to 100 from 20 different IP addresses? Hence, once you start the IP rotation, you should also use a package such as a fake user agent (`https://pypi.org/project/fake-useragent/`) or Scrapy user agents (`https://pypi.org/project/Scrapy-UserAgents/`) to generate new user agent strings as shown in Listing 8-17.

Listing 8-17. Randomly generated user agent

```
from fake_useragent import UserAgent
ua = UserAgent()
print(ua.random)
#Output
Mozilla/5.0 (Windows NT 6.4; WOW64) AppleWebKit/537.36 (KHTML, like Gecko)
Chrome/41.0.2225.0 Safari/537.36
```

If you are a website owner and maintain web logs, then I suggest that you scrape it to extract user agents of real visitors to your website, load it to a database, and use that for the user-agent rotation since that is a more robust approach than relying on external packages which in turn have to scrape it from some other website such as useragentstring.com (http://useragentstring.com/pages/useragentstring.php?name=Chrome).

Cloudflare

Cloudflare is an extremely popular content delivery network used by over 20% of the top 1 million sites according to builtwith.com. Their free plan includes protection against distributed denial-of-service attacks (DDoS). Let us recall the discussion we had in Chapter 2 about how a distributed crawler hitting a particular web domain without timeouts appears very similar to a DDoS-type attack. This is the reason why we should have reasonable timeouts between requests, but sometimes that's not enough to prevent getting flagged as a potential DDoS attack by Cloudflare.

It serves an "under attack mode" page whenever it suspects that the request is not legitimate; the page looks similar to the one shown in its documentation page (https://support.cloudflare.com/hc/en-us/articles/200170076-Understanding-Cloudflare-Under-Attack-mode-advanced-DDOS-protection-).

A Cloudflare under attack page is a client-side antibot measure, and it checks if JavaScript is enabled and issues a challenge based on it. This should be easy enough to bypass if you are using a real browser like we do in Selenium-based web scraping; but more commonly, we are requesting the HTML directly, and that will get caught by this page, leading to an issuance of a CAPTCHA or, worse still, putting out an IP address on a blacklist.

One of the ways to bypass this is using modules such as cfscrape (`https://pypi.org/project/cfscrape/`), cloudscraper (`https://pypi.org/project/cloudscraper/`), and so on. It tries to impersonate a browser so that we can bypass this page and go on to scrape the web page.

These packages will work to some extent, and hence I mentioned it here, but for sustained workloads, you will only get good results by running a real browser through Selenium even if the content you need is not dependent on running JavaScript. Cloudflare will serve a type of CAPTCHA called hCaptcha after showing the under attack page in case you are still exceeding a request rate threshold or trying to access a page or some other behavior specifically flagged by a website in firewall rules (`https://blog.cloudflare.com/moving-from-recaptcha-to-hcaptcha/`). In these cases, you will anyways need to have a real browser running via Selenium for using CAPTCHA solving services discussed in the following section.

CAPTCHA solving services

CAPTCHA is one of the most potent antibot challenge test a website can deploy to prevent or at least slow down scraping activities short of asking the users to log in for gaining access to the content.

There are a lot of ways of bypassing traditional image-based CAPTCHA by image recognition AI models. However, this changed when Google rolled out reCAPTCHA v2 and successor versions which have been proven pretty hard to bypass using automated methods except some intermittent successes such as the unCaptcha project from a team in the University of Maryland (`https://uncaptcha.cs.umd.edu/`). The only reliable method of solving it now is by calling CAPTCHA solving services such as `https://anti-captcha.com`, `www.deathbycaptcha.com/`, and `www.solverecaptcha.com/apidoc/` that charge about $.001–.004 per solved CAPTCHA by relying on low-cost manual labor based in mainly low-income countries.

It's pretty easy to understand the implementation side of reCAPTCHA; pages containing it have a div tag with a site key; you have to extract its value and send over the value and page URL to a CAPTCHA solving service API provider which will return back the recaptcha response.

```
<div id="g-recaptcha" class="g-recaptcha" data-sitekey="your_site_key">
```

Listing 8-18 shows a pseudocode of calling one of these services for Google's reCAPTCHA; the exact code will depend on the service you use and the type of page you are being served a CAPTCHA. In this example, I have assumed that we are on a form where we have to pass a reCAPTCHA check before hitting submit. This is similar to querying for the domain information on the Ahrefs backlink checker page we already saw in Figure 1-6, Chapter 1. Once you get back the response code from one of these providers, all you have to do is use the execute_script method of the Selenium driver to execute it and perform some action on the page like the submit button to get the page you want to scrape. It takes 10–30 seconds to get a response back since it is human solved so crawling speed is slowed down considerably if you have to solve a reCAPTCHA for every scraped web page.

Listing 8-18. Pseudocode for a reCAPTCHA solving service

```
from selenium import webdriver
browser = webdriver.Chrome

captcha_site_key = browser.find_element_by_class_name('g-recaptcha').get_
attribute('data-sitekey')
#...(call CAPTCHA solving service API with site key and url
# It will return back  g_response_code
js_code = 'document.getElementById("g-recaptcha-response").innerHTML =
"{}";'.format(g_response_code)

browser.execute_script(js_code)
# Now perform whatever action you need to do on the page like hitting a
submit button
browser.find_element_by_tag_name('form').submit()
```

The div tag attributes and the code earlier vary a little with different variants of CAPTCHA such as hCaptcha used by Cloudflare, but you can get detailed example codes on documentation pages of CAPTCHA solving API services.

Summary

We learned about a production-ready crawling framework called Scrapy and went through examples of using it to upload raw web crawls to an S3 bucket. Later, we discussed advanced crawling strategies using proxy IP rotation, user-agent rotation, and CAPTCHA solving techniques.

I will end this book on the same note I began in Chapter 1 by reiterating that you need to pick and choose your battles, and web crawling is such a large field that you will probably not be able to do everything in house. I also think lots of startups make this mistake when building a web crawling pipeline instead of specializing in a few core areas. It's much better to focus initially on aspects that contribute the most to their products' intellectual property (IP) and leave the rest to third-party providers at least for powering the proof-of-concept products. Many managers tend to overlook the cost of maintaining a web scraping/crawling pipeline by underestimating the fragile nature of most web scrapers and constantly need to update the scraper to match with future website changes.

My second recommendation to new practitioners is avoid focusing too much on libraries such as SQLAlchemy or Scrapy with high-level abstractions when first starting out and spend some time understanding the low-level implementation details and ways you can make the underlying code more efficient. In a similar vein, whenever possible, try and use the more efficient libraries written in C such as lxml, regex engines such as re2, and so on over more slower pure Python-based variants. I hope by using this book you now have a good enough idea to understand the underlying principles of web crawling in a framework-agnostic way.

Lastly, I think it's imperative that all software developers in the web crawling space pick up some machine learning and natural language processing (NLP) skills. I had a wide-ranging topic to cover in this book, and hence I couldn't cover advanced vectorization methods such as BERT in Chapter 4, but advancements such as these have become mainstream and both Google and Bing are using them now to power their search queries. There have been drastic improvements in all NLP areas, and these will continue to open up new use cases and ways we process web crawl data in days to come.

Index

A

Agglomerative clustering, 209
Ahrefs, 6, 9, 10, 12, 278, 315
AJAX, 74, 75
Alexa, 11–13, 278
Alternative Financial Datasets, 15–17
Amazon Athena, 325–330, 335, 370
Amazon Elastic Compute
    Cloud (Amazon EC2), 88, 110
Amazon machine images (AMI), 112
Amazon Relational Database Service
    (RDS), 88, 242, 243
Amazon simple notification service (SNS),
    88, 124, 125, 127, 129
Amazon web services (AWS), 85, 87–93,
    98–101, 110–113, 115, 124, 125,
    127, 129, 133

B

Backlinks, 8–13
Backlinks database, 315, 317–323
Beautiful Soup, 31, 37, 39, 40, 42, 43, 47,
    49, 51, 52, 67, 76, 78
Boto3, 86, 101, 107, 125

C

CAPTCHA, 383, 386, 388, 390–392
CAPTCHA solving services, 371, 391, 392

C

Cascading Style Sheets (CSS), 33–35, 57, 84
Cloud computing, 85–87, 124, 133
Coherence, 192, 193, 196, 199, 201
Common Crawl Foundation, 277, 290,
    300, 324
Common crawl index, 283–287, 289, 290,
    326, 334–338
Crawl delay, 371, 373
Crawl depth, 61, 62, 373
Crawl order, 61, 62, 373
Cyberduck, 107–109, 122

D

Database schema, 229, 231, 232, 234, 235,
    239, 275
Data definition language (DDL), 231, 242,
    244, 249, 254, 255, 326
Data Manipulation Language
    (DML), 252, 254, 255, 257, 272
Data Query language (DQL), 252, 254,
    259, 264, 272, 275
Dbeaver, 226, 239, 241
Distributed computing, 350, 358, 360,
    362–369
Document object model (DOM), 47, 72, 74
Domain ranking, 325–332, 334, 348

E

Exploratory data analytics (EDA), 162–164

395

Printed in the United States
By Bookmasters